Research Strategies for the U.S. Global Change Research Program

CHRIS RAPLEY

Committee on Global Change
(U.S. National Committee for the IGBP)
of the
Commission on Geosciences, Environment, and Resources
National Research Council

NATIONAL ACADEMY PRESS
Washington, D.C. 1990

NOTICE: The project that is the subject of this report was approved by the Governing Board of the National Research Council, whose members are drawn from the councils of the National Academy of Sciences, the National Academy of Engineering, and the Institute of Medicine. The members of the committee responsible for the report were chosen for their special competences and with regard for appropriate balance.

This report has been reviewed by a group other than the authors according to procedures approved by a Report Review Committee consisting of members of the National Academy of Sciences, the National Academy of Engineering, and the Institute of Medicine.

The National Academy of Sciences is a private, nonprofit, self-perpetuating society of distinguished scholars engaged in scientific and engineering research, dedicated to the furtherance of science and technology and to their use for the general welfare. Upon the authority of the charter granted to it by the Congress in 1863, the Academy has a mandate that requires it to advise the federal government on scientific and technical matters. Dr. Frank Press is president of the National Academy of Sciences.

The National Academy of Engineering was established in 1964, under the charter of the National Academy of Sciences, as a parallel organization of outstanding engineers. It is autonomous in its administration and in the selection of its members, sharing with the National Academy of Sciences the responsibility for advising the federal government. The National Academy of Engineering also sponsors engineering programs aimed at meeting national needs, encourages education and research, and recognizes the superior achievements of engineers. Dr. Robert M. White is president of the National Academy of Engineering.

The Institute of Medicine was established in 1970 by the National Academy of Sciences to secure the services of eminent members of appropriate professions in the examination of policy matters pertaining to the health of the public. The Institute acts under the responsibility given to the National Academy of Sciences by its congressional charter to be an adviser to the federal government and, upon its own initiative, to identify issues of medical care, research, and education. Dr. Samuel O. Thier is president of the Institute of Medicine.

The National Research Council was organized by the National Academy of Sciences in 1916 to associate the broad community of science and technology with the Academy's purposes of furthering knowledge and advising the federal government. Functioning in accordance with general policies determined by the Academy, the Council has become the principal operating agency of both the National Academy of Sciences and the National Academy of Engineering in providing services to the government, the public, and the scientific and engineering communities. The Council is administered jointly by both Academies and the Institute of Medicine. Dr. Frank Press and Dr. Robert M. White are chairman and vice chairman, respectively, of the National Research Council.

This work was sponsored by the National Science Foundation, National Aeronautics and Space Administration, National Oceanic and Atmospheric Administration, United States Geological Survey, United States Department of Agriculture, Office of Naval Research, and Department of Energy. Contract No. OCE 8713699.

Library of Congress Catalog Card No. 90-62876
International Standard Book Number 0-309-04348-4

Additional copies of this report are available from:

National Academy Press
2101 Constitution Avenue, NW
Washington, DC 20418

S191

Printed in the United States of America

Cover art by David Chang

Committee on Global Change
(U.S. National Committee for the IGBP)

HAROLD MOONEY, Stanford University, *Chairman*
FRANCIS P. BRETHERTON, University of Wisconsin
D. JAMES BAKER, JR., Joint Oceanographic Institutions, Inc.
KEVIN C. BURKE, National Research Council
WILLIAM C. CLARK, Harvard University
MARGARET B. DAVIS, University of Minnesota
ROBERT E. DICKINSON, University of Arizona
JOHN IMBRIE, Brown University
ROBERT W. KATES, Brown University
THOMAS F. MALONE, St. Joseph College
MICHAEL B. McELROY, Harvard University
BERRIEN S. MOORE III, University of New Hampshire
ELLEN S. MOSLEY-THOMPSON, Ohio State University
PAUL G. RISSER, University of New Mexico
PIERS J. SELLERS, University of Maryland

Ex-Officio Members (U.S. Members, ICSU Special Committee for the IGBP)

JOHN A. EDDY, University Corporation for Atmospheric Research
JAMES J. McCARTHY, Harvard University
S. ICHTIAQUE RASOOL, National Aeronautics and Space Administration

Staff

JOHN S. PERRY, Staff Director
RUTH DeFRIES, Senior Program Officer
CLAUDETTE BAYLOR-FLEMING, Senior Program Associate

Commission on Geosciences, Environment, and Resources

M. GORDON WOLMAN, The Johns Hopkins University, *Chairman*
ROBERT C. BEARDSLEY, Woods Hole Oceanographic Institution
B. CLARK BURCHFIEL, Massachusetts Institute of Technology
RALPH J. CICERONE, University of California at Irvine
PETER S. EAGLESON, Massachusetts Institute of Technology
GENE E. LIKENS, New York Botanical Gardens
SCOTT M. MATHESON, Parsons, Behle & Latimer
JACK E. OLIVER, Cornell University
PHILIP A. PALMER, E.I. du Pont de Nemours & Co.
FRANK L. PARKER, Vanderbilt University
DUNCAN T. PATTEN, Arizona State University
MAXINE L. SAVITZ, Allied Signal Aerospace Company
LARRY L. SMARR, University of Illinois, Urbana-Champaign
STEVEN M. STANLEY, Case Western Reserve University
SIR CRISPIN TICKELL, Radcliffe Observatory
KARL K. TUREKIAN, Yale University
IRVIN L. WHITE, N.Y. State Energy Research and Development Authority
JAMES H. ZUMBERGE, University of Southern California

STEPHEN RATTIEN, Executive Director
STEPHEN D. PARKER, Associate Executive Director
JANICE GREENE, Assistant Executive Director
JEANETTE SPOON, Administrative Associate
GAYLENE DUMOUCHEL, Administrative Assistant

Preface

This report is a step in the evolving process of defining the scientific needs for understanding changes in the global environment, changes that are of great concern to the public, policymakers, and the international scientific community. Following the National Research Council's 1983 report *Toward an International Geosphere-Biosphere Program: A Study of Global Change*, 1986 report *Global Change in the Geosphere-Biosphere: Initial Priorities for an IGBP*, and 1988 report *Toward an Understanding of Global Change: Initial Priorities for the International Geosphere-Biosphere Program*, this report builds and expands on the scientific priorities established by the scientific community over the past 7 years. The report recommends research strategies to address the priorities identified in the 1988 report for implementation in the U.S. Global Change Research Program (USGCRP), the national program that contributes to the goals of the international research programs addressing global change.

The report was developed through a number of working groups established by the Committee on Global Change, each under the chairmanship of a member of the committee. Over an 18-month period in 1988 and 1989, these working groups addressed their respective charges to develop scientific approaches and research strategies for each of the five priorities identified in the committee's 1988 report (water-energy-vegetation interactions, fluxes of materials between terrestrial ecosystems and the atmosphere and oceans, biogeochemical dynamics of ocean interactions with climate, earth system history and modelling, and human interactions with global change). The reports from these working groups were used by the committee as inputs to the five chapters on focused studies in this report. In addition to participating in these working groups, members of the committee, working with others in the scientific community, highlighted approaches to developing integrated earth system models and documenting global change over the long term. The results of these efforts formed the basis of chapters 2 and 8 of this report.

Recognizing the evolving nature of global change research, the committee intends that the topics recommended for focused studies within the USGCRP be viewed as initial priorities. They are not intended to span the range of issues that need to be addressed. Rather, they are based on the committee's analysis in its 1988 report of the most critical gaps, not being addressed by existing programs, in the scientific knowledge needed to understand the changes that are occurring in the earth system on time scales of decades to centuries. For example, issues related to the physical climate system that are currently being addressed through the World Climate Research Program are not highlighted in this report. Nor does the committee intend that the topics selected for priority attention span the range of all global change research that needs to be carried out within more traditional disciplinary frameworks. The integration of many aspects of biology, such as the loss of biodiversity and its effects on ecosystem structure and functioning, is still an important object for research and for the development of appropriate models. Within the social sciences, in addition, research needs are currently being defined by the NRC Committee on the Human Dimensions of Global Change.

Many people were involved in the development of this report. First and foremost, my fellow members of the Committee on Global Change worked long and hard. The members of the working groups and the liaison representatives from the interagency Committee on Earth and Environmental Sciences, too numerous to name here but indicated in each of the chapters, provided the expertise and energy required to develop their reports. John Perry and Ruth DeFries of the National Research Council staff deftly managed the critical task of turning drafts into a coherent report, and Roseanne Price provided outstanding editorial services. The contributions of each of these people are gratefully acknowledged.

Harold A. Mooney, Chairman
Committee on Global Change

Contents

SUMMARY OF RECOMMENDATIONS 1
   Integrated Modeling of the Earth System, 2
   Focused Studies to Improve Our Understanding of Global Change, 3
      Earth System History and Modeling, 3
      Human Sources of Global Change, 4
      Water-Energy-Vegetation Interactions, 5
      Terrestrial Trace Gas and Nutrient Fluxes, 5
      Biogeochemical Dynamics in the Ocean, 6
   Documenting Global Change, 7
   References, 8

1 INTRODUCTION 9
   The U.S. Global Change Research Program, 10
   International Programs: IGBP and WCRP, 11
   Objectives and Organization of This Report, 12
   References, 15

2 INTEGRATED MODELING OF THE EARTH SYSTEM 16
   Overview, 16
   Atmosphere-Terrestrial Subsystem, 23
      A Modeling Strategy: Prognosis for Progress, 25
      Research Priorities, 32
      Summary, 35
   Physical-Chemical Interactions in the Atmosphere, 36
      A Modeling Strategy: Prognosis for Progress, 41
      Research Priorities, 42
      Summary, 45

Atmosphere-Ocean Subsystem, 46
    Uncoupled Models, 47
    Coupled Models, 49
    A Modeling Strategy: Prognosis for Progress, 50
    Research Priorities, 55
    Summary, 58
Critical Model Tests, 59
    The Challenge and Critical Tests, 59
    The Interface Models, 60
Infrastructure, 61
References, 63

3 EARTH SYSTEM HISTORY AND MODELING 67
Overview, 67
    Contribution of Geologic Studies to Global Change, 67
    Specific Research Initiatives, 69
    Priorities, 70
    Themes of the Proposed Research, 70
    Implementation of the Research Plan, 70
Holocene High-Resolution Environmental Reconstructions, 71
    The Last 1,000 to 2,000 Years, 71
    Earlier Holocene Millennial-Scale Fluctuations, 77
Glacial-Interglacial Cycles, 77
    The Last 40,000 Years, 80
    The Last Glacial Cycle (Last 130,000 Years), 85
    The Last Few Glacial-Interglacial Cycles (Last 500,000 Years), 91
System Responses to Large Changes in Forcing, 92
    Environments of Extreme Warm Periods, 93
    Climate-Biosphere Connections During Abrupt Changes, 95
Critical Program Elements, 98
    Sample Acquisition, 98
    Environmental Calibration, 99
    Correlation of Records, 100
    Data Management, 101
International Cooperation, 101
References, 102

4 HUMAN SOURCES OF GLOBAL CHANGE 108
Overview, 108
    Background, 109
    Priority Recommendations, 110
The Research Program, 111

Industrial Metabolism, 112
    Integration and Synthesis, 112
    Process Studies and New Data, 113
Land Transformations, 117
    Integration and Synthesis, 118
    Process Studies, 120
    Data Needs, 122
Integrative Studies Across Land Use and Industry, 123
    Global Model of Greenhouse Gas Emissions, 124
    Earth Systems Information Flow Diagram for Human Interactions, 124
    Driving Forces: Population, Economy, Technology, and Institutions, 125
Implementation Requirements, 125
    Related Institutional Efforts on Human Interactions with Global Change, 125
    Investigator-Initiated Research, 126
    Education and Training, 127
    Data Preparation and Dissemination, 127
The Steps Beyond, 127
Notes, 128
References, 129

5 WATER-ENERGY-VEGETATION INTERACTIONS 131
Overview, 131
Data Needs and Experiments, 133
    Global Data Needs, 136
    Experiments, 144
    Fundamental Research and Laboratory Work, 148
Modeling, 152
    Intermodel Transfer Packages, 155
    Phenological Descriptions for LSPs, 156
    Hydrological Models, 156
    Surface/Planetary Boundary Layer Models, 157
    Ecosystem Structure Models, 157
    Radiative Transfer/Plant Physiology Models, 158
    Soil Genesis Models, 160
    Sensitivity Analyses, 160
    Summary, 161
Infrastructure, 161
    Operational Observations, 161
    Satellite Data Processing, 161

Centers for Research and Monitoring, 162
Education, 162
Interagency and International Coordination, 162
References, 162

6 TERRESTRIAL TRACE GAS AND NUTRIENT FLUXES........ 164
Overview, 164
Problem Definition, 165
General Approach, 166
Research Needs, 169
Trace Gases, 169
Nutrient and Material Fluxes, 182
Methods and Instruments, 188
Models, 188
Instrumentation for Measuring Fluxes, 191
Cross-Cutting Issues, 192
References, 194

7 BIOGEOCHEMICAL DYNAMICS IN THE OCEAN............. 200
Overview, 200
Status of Existing Efforts, 202
Biogeochemical Fluxes, 202
Ocean-Atmosphere Interface, 204
Oceanic Ecosystem Response to Climatic Change, 205
Physical Processes, 206
Polar Processes, 209
Status of Modeling and Monitoring Efforts, 210
The Need for Modeling, 210
The Need for Monitoring, 210
Recommendations for Enhanced Support, New
  Initiatives, and Research Programs, 212
References, 214

8 DOCUMENTING GLOBAL CHANGE......................... 215
Overview, 215
Measurement Strategy, 216
Monitoring Requirements, 216
Global Synthesis, 217
Process Studies, 221
Existing and Planned Observing Systems, 224
International Coordination, 232
Information and Data Management, 233
Data System Requirements, 233

Kinds of Data Needed, 234
Functions of a Data and Information System, 234
Creating a New System, 235
References, 239

APPENDIXES
A List of Participants in the Workshop on Human Interactions with Global Change, 243
B A Selective Literature Review on the Human Sources of Global Environmental Change, *by Vicki Norberg-Bohm,* 246
C Related Institutional Efforts on Human Interactions with Global Change, 285

Summary of Recommendations

There is no better example of the need for scientists and decision makers to work together than the broad range of issues concerning global environmental change, including greenhouse warming, stratospheric ozone depletion, tropical deforestation, and other changes not yet identified or predicted. Policies to slow emissions or influence human behavior to mitigate or adapt to change must be based on scientific assessments of how the global environment will change in the future. The scientific information, however, is in many cases uncertain and incomplete, owing to our incomplete understanding of the earth system and its interacting components. Substantial efforts on the part of the scientific community and federal science agencies will be required over the coming decades to improve our ability to understand the earth system and hence to project future changes in the global environment.

With this perspective the U.S. Global Change Research Program (USGCRP) was initiated in FY 1990 as the federal government's effort to "establish the scientific basis for national and international policy-making relating to national and human-induced changes in the global change earth system" (CES, 1990). The objectives of the program are to establish an integrated, comprehensive, long-term program of documenting the earth system on a global scale; develop integrated conceptual and predictive earth system models; and conduct a program of focused studies to improve understanding of the physical, geological, chemical, biological, and social processes that influence earth system processes and trends on global and regional scales. The USGCRP embodies the U.S. scientific contributions to the two principal international research programs concerning global change, the International Geosphere-Biosphere Program (IGBP) of the International Council of Scientific Unions (ICSU), and the World Climate Research Program (WCRP) organized jointly under ICSU and the World Meteorological Organization. The former is addressing the interactive biological, physical, and chemical processes that regulate the earth system, and the latter is addressing the physically dynamic and radiative aspects of the climate system.

This report recommends specific initiatives required to achieve the objectives of the USGCRP to (1) document long-term changes in the earth system on a global scale and collect data necessary for process studies and modeling, (2) develop integrated models of the earth system, and (3) improve understanding of global change through focused studies of earth system processes. The initiatives discussed in this report elaborate on previous recommendations from the Committee on Global Change on initial priorities for U.S. contributions to the IGBP (NRC, 1988). The recommendations are summarized below.

INTEGRATED MODELING OF THE EARTH SYSTEM

In order to develop a fully coupled, dynamical model of the earth system useful for projecting changes over a multidecadal time scale, a step-by-step buildup in complexity is required. The multidecadal time frame demands that changes in the biosphere and biogeochemical feedbacks, which happen over similar lengths of time, be incorporated into earth system models. For the near term, progress will be achieved through linking previously unlinked components of the earth system (e.g., coupling terrestrial systems with the atmosphere) or adding specific subsystems to existing models (e.g., coupling oceanic-atmospheric general circulation models (GCMs) or adding a marine biospheric model to an oceanic GCM).

The committee recommends a focus on three critical subsystems, each of which represents an interface between various components of the earth system:

- *Models that couple the atmosphere-terrestrial subsystem.* Modeling these interactions requires coupling successional models to biogeochemical models to physiological models. The models must address how global environmental changes, including the effects of land use and chemical stress, affect terrestrial ecosystems and how ecosystem changes affect the global system. A primary challenge to be addressed is how processes with vastly differing rates of change, from photosynthesis to community change, are coupled to each other and to the atmosphere. Process studies and associated modeling activities are required to develop a better understanding of these processes, as discussed in chapters 5 and 6.

- *Models that couple the physics and the chemistry of the atmosphere.* Progress in modeling these interactions requires a better understanding of the sources of trace gases, chemical processes and reaction channels in gas and aqueous phases, and transport processes by advection and convection. As with models that couple the atmosphere-terrestrial subsystem, progress in modeling depends on improvements in understanding of fundamental physical and chemical processes and the nature of their coupling.

- *Models that couple the atmosphere and the ocean.* Particularly important tasks are the scaling of the biological-biogeochemical components from local-regional domains to basin-global domains, formation of the upper mixed-layer physics, and possible biological feedbacks on mixed-layer dynamics such as shading due to phytoplankton blooms. The development of models should be encouraged along two parallel paths: one to develop basin- and global-scale models with increasing levels of coupling and a second to develop a series of regional fine-scale models that could provide boundary conditions and parameterization tests for the larger-scale models. As discussed in chapter 7, existing and planned field programs offer valuable opportunities for improving fundamental understanding of oceanic processes and hence for improving modeling capabilities.

Focusing on these three subsystems will not only provide the building blocks for fully integrated earth system models but also establish prototypes for the development of larger models. Even failure in these efforts can provide valuable lessons to the scientific community. Early tests of prototypical earth system models, including tests using the record of the past (see chapter 3), should begin as soon as possible.

FOCUSED STUDIES TO IMPROVE OUR UNDERSTANDING OF GLOBAL CHANGE

The committee recommends that the following five initiatives be undertaken as focused studies to improve our understanding of global change. The initiatives are not intended to provide a comprehensive view of all the required studies within the USGCRP. Rather, they are contributions to the full suite of research needs covering a broad spectrum including the physical climate, solid earth processes, and solar influences.

Each of these initiatives requires a combination of process studies, modeling activities of the particular components of the system addressed by the initiative, and observations to collect pertinent data. The efforts to document and develop integrated models of the earth system support and are supported by these initiatives, but each initiative also has distinct data and modeling requirements.

Earth System History and Modeling

Earth history contains a rich record unique in its potential to provide comprehensive case studies in global change. The coupled response of climate, hydrology, biogeochemical cycles, and the biosphere to a large number of perturbations, including more recent human activities, is preserved in these records. Understanding these linkages is essential for accurately predicting the future evolution of climate and the biosphere. Geologic observations also present the only opportunity to verify independently models of how the earth system operates.

The committee recommends that the following activities receive priority within the earth system history initiative:

• establishment of a global network of high-resolution climate histories for the Holocene (last 10,000 years) with emphasis on the last 1,000 to 2,000 years. These records will provide a frame of reference for comparison with any future warming due to greenhouse gases.

• research to understand the origin of glacial-interglacial fluctuations. This research will provide an excellent opportunity to examine system response to known forcing. Special emphasis should be on (1) abrupt changes, (2) the carbon cycle, (3) tropical environments at the last glacial maximum, and (4) coupling of the different components of the climate system.

• evaluation of system response to large changes in boundary conditions. These studies will provide a sturdy test of climate models and enable evaluation of the linkages between climatic change and the biosphere. Research will focus on reconstruction of the environment of the Pliocene warm interval (3 to 5 million years ago) and evaluation of the climate-biosphere connection during the Eocene-Oligocene (30 to 40 million years ago) transition.

Human Sources of Global Change

In order to gain a systematic understanding of how human activities alter the global environment, particularly through changes in land use and industrial metabolism, the committee recommends that

• integrative models of human sources of industrial emissions and land cover change be developed, particularly global models of how human activities relating to agriculture cause changes in chemical fluxes and land cover, global models of changes in greenhouse gas emissions from human sources, and regional models of how human activities cause land cover conversions of tropical forests and wetlands.

• studies be initiated on the social and economic forces driving the following processes important for global change: fertilization in agriculture, biomass burning, and intensity of energy and material use. These studies will provide the understanding important for development of the integrative models noted above.

• data critical for model development and process studies be collected and organized with priority given to (1) a global data base allowing comparisons of population and land cover, land use, and land capability data at a census district level and (2) a global data base on historical energy and materials use across a range of human activities.

Water-Energy-Vegetation Interactions

The goal of this research initiative is to develop the underlying understanding and models of the interactions between the land biota and atmosphere, with particular reference to the exchanges of energy, water, heat, and trace gases. The process studies and models should explore the consequences of global change on terrestrial ecosystems and, ultimately, should be adequate to predict the response of the climate system to perturbations in ecosystems. The committee recommends that

- global monitoring be adequate to compile data sets for those parameters necessary to calculate land surface-atmosphere fluxes of radiation, water, heat, and carbon dioxide. These parameters are vegetation index; cloudiness and precipitation; temperature, humidity, and wind speed; surface temperature; runoff; and soil type.
- process studies and field campaigns be initiated to test models relating biophysical and biochemical states to surface fluxes and remotely sensed variables, including models that link surface states and radiances, models that relate water use efficiency and concentrations of carbon and nutrients, and models that relate ecosystem structure and function and ecosystem response to global changes.
- modeling initiatives be implemented including development of coupled land surface parameterizations (LSPs) and general circulation models (GCMs) for sensitivity studies of land use change and direct and indirect effects of increases in carbon dioxide; models describing fluxes; and ecophysiological models of carbon-water relations.

Terrestrial Trace Gas and Nutrient Fluxes

This initiative aims to (1) improve understanding of the ecosystem processes most important for determining the fluxes of radiatively active gases between the land and the atmosphere, in order to predict how changes in climate and land use alter gas emissions and (2) improve our understanding of the effects of land use changes on nutrient transfer to river, estuarine, and ocean systems, and especially of consequent feedbacks to climate through, for example, long-term changes in oceanic productivity. The committee recommends

- measurements and models describing how changes in climate and ecosystems, precipitation, and land use alter carbon storage in ecosystems.
- measurements and improved models for the major sources and sinks

for carbon dioxide and the global pattern of carbon dioxide transport in the atmosphere.

- process studies that relate methane production, consumption, and fluxes to environmental parameters and to changes in ecosystem structure and function.
- measurements of biogenic sulfur emissions (hydrogen sulfide, dimethylsulfide, carbonyldisulfide, and methyl mercaptan) from a variety of terrestrial ecosystems, especially those likely to be significant regional and global sources, e.g., wet tropical regions, coastal marshes, boreal forest peatlands, tundra bogs, rice paddies, landfills, and industry.
- improvement of the data base describing ozone deposition in various ecosystems and the responses of ecosystems to acid deposition.
- measurement and modeling of the fate of nutrients, pollutants, and sediments from terrestrial systems in streams and the oceans as affected by land use changes.
- microscale and mesoscale models for extrapolating specific processes, such as trace gas source and sinks and fluxes, to regional and global scales.

Biogeochemical Dynamics in the Ocean

The objective of this initiative is to develop the capability to predict the effects of climatic change on the ocean's physical, chemical, and biogeochemical processes, especially as they feed back to climate via the release or absorption of radiatively important gases such as carbon dioxide and organic sulfur species. To meet this objective, the committee recommends the following:

- U.S. support for the implementation phases of established projects relevant to understanding the role of the ocean in global change: the Joint Global Ocean Flux Study (JGOFS), International Global Atmospheric Chemistry (IGAC), Tropical Ocean and Global Atmosphere (TOGA), and Global Energy and Water Cycle Experiment (GEWEX), including support for in situ and satellite observations, modeling, and linkages among these projects.
- initiation of long-term systematic measurements of relevant oceanic properties and processes including dissolved carbon dioxide, ocean color, surface wind and currents, and rainfall with in situ and satellite techniques.
- initiation of coastal ocean studies to understand (1) the transfer of materials (nutrients and pollutants) from land to ocean and their sensitivity to global change (e.g., land use and changes in hydrological cycles and sea level) and (2) the dynamics of ecosystems, and how they can be influenced by global change, especially those ecosystem components particularly relevant to biogeochemical cycles and of particular commercial interest.

SUMMARY OF RECOMMENDATIONS 7

DOCUMENTING GLOBAL CHANGE

The systems established to monitor global change over the coming decades, to provide data for process studies and modeling aimed toward improving our understanding of global change, and to manage the data and information so that they are accessible to researchers are crucial to the success of the USGCRP.

Long-term global-scale monitoring needs are of two kinds: (1) to quantify the magnitude of the driving forces that may bring a long-term change in the state of the earth system, for example, solar irradiance, trace gas concentrations, and land use change, and (2) to monitor the state variables or "vital signs" of the earth where such changes are liable to manifest, for example, tropospheric, stratospheric, and surface temperature, ocean uptake of carbon dioxide, sea level, and soil and vegetation characteristics. Relative priorities for monitoring individual variables must be judged not only in terms of their contribution to the monitoring requirements but also in terms of their importance for validating models and advancing our understanding of specific processes and the readiness of the observation and analysis techniques involved.

Globally synthesized products derived from measurements—for example, a global data set for latent heat flux derived from data on radiation balance, temperature, moisture, and field studies—will be required to achieve the modeling objectives described in chapter 2. Data needs in relation to studies of specific processes as discussed in chapters 3 through 7 of this report place additional requirements on the observation and monitoring program.

In order to meet the data requirements for the USGCRP, satellite systems, large-scale field studies, and surface networks are needed. The current international operational satellite system, satellite research missions, and a total system such as the Earth Observing System together provide the basis for the space-based portion of a global change observing system. Several surface observation networks such as the Background Air Pollution Monitoring Network (BAPMON) and networks organized by the World Meteorological Organization and the U.N. Environment Program contribute to the surface-based observational requirements. Existing ecological research sites such as the National Science Foundation's Long-Term Ecological Research Sites, as well as the planned IGBP Regional Research Centers, could also contribute. International coordination and data exchange, already fairly extensive for space-based programs, need to be strengthened for an effective overall global change observational strategy.

An important concern with space-based observations is the discontinuity of key measurements such as global stratospheric ozone levels, the earth's radiation budget, and the biological productivity of oceans. In addition, research missions are needed to test technologies for direct satellite measurement of precipitation and soil moisture.

The USGCRP will make unprecedented demands for the assembly and dissemination of large volumes of diverse data and information. Data and information must be accessible to researchers at the lowest cost possible, and the system must involve the scientific user community at all stages of development and operation. In addition, data documentation must be able to pass the "20-year test"; i.e., 20 years from now, someone unfamiliar with the data should be able to fully understand and use the data solely with the aid of the documentation archived with the data set.

REFERENCES

Committee on Earth Sciences (CES). 1990. Our Changing Planet: The FY 1991 U.S. Global Change Research Program. Federal Coordinating Council for Science, Engineering, and Technology. Office of Science and Technology Policy, Washington, D.C.

National Research Council (NRC). 1988. Toward an Understanding of Global Change: Initial Priorities for U.S. Contributions to the International Geosphere-Biosphere Program. National Academy Press, Washington, D.C.

1

Introduction

In the past few years, global environmental change has increasingly taken a central place on the stage of national and international policy discussions. In many cases, however, the scientific information that forms the basis of discussions about far-reaching policies to mitigate or adapt to change is uncertain and incomplete. Decisions on energy policies to curtail emissions of greenhouse gases, for example, or decisions to expend funds to adapt to possible changes in the global environment, must rest on a scientific understanding of the earth system that is only beginning to emerge.

Global environmental change encompasses many facets. The environment is undergoing significant alterations as a result of human activities superimposed on the natural variability of the earth system. The "greenhouse effect" and consequent climatic change are currently of great concern to the public. Other changes in the global environment include depletion of stratospheric ozone, deforestation, acid deposition, as well as other as yet undetected and unanticipated changes. The scientific understanding of each of these manifestations of global change hinges on the ability to understand the total earth system and the interactions between the atmosphere, oceans, land, and biota—including humans.

While global change has been the topic of high-level policy discussions, it has also become the focus of increasingly active planning for scientific programs. Internationally, the International Geosphere-Biosphere Program (IGBP) and the ongoing World Climate Research Program (WCRP) constitute closely interrelated and mutually complementary efforts focusing, respectively, on the biological-chemical and the physical-dynamic aspects of long-term global environmental change. The U.S. Global Change Research Program (USGCRP) coordinates the U.S. contributions to these international programs. Together, these efforts aim toward understanding the earth system in order to improve our predictive capabilities of global environmental change and thus to provide a foundation for sound policy decisions.

THE U.S. GLOBAL CHANGE RESEARCH PROGRAM

The U.S. Global Change Research Program, initiated in fiscal year 1990, represents the U.S. federal government's scientific effort to understand, monitor, and predict global change. The USGCRP is an interdisciplinary program coordinated across many federal agencies and is designed to (CES, 1989a,b, 1990)

> gain an adequate predictive understanding of the interactive physical, geological, chemical, biological, and social processes that regulate the total earth system and, hence establish the scientific basis for national and international policy formulation and decisions relating natural and human-induced changes in the global environment and their regional impacts.

It operates under the Federal Coordinating Council for Science, Engineering, and Technology (FCCSET) Committee on Earth and Environmental Sciences (formerly the Committee on Earth Sciences) with the advice of the Committee on Global Change and other units of the National Research Council.

The overarching scientific objectives of the program are to establish an integrated, comprehensive long-term program of documenting the earth system on a global scale; conduct a program of focused studies to improve our understanding of the physical, geological, chemical, biological, and social processes that influence earth system processes and trends on global and regional scales; and develop integrated conceptual and predictive earth system models. To meet these objectives, the plan for the USGCRP currently embodies seven interdisciplinary science elements (CES, 1989b):

• Climate and hydrological systems. The examination of the physical processes that govern physical climate and the hydrological cycle, including interactions between the atmosphere, hydrosphere (i.e., oceans, surface and ground water, clouds, and so on), cryosphere, land surface, and biosphere.

• Biogeochemical dynamics. The study of the sources, sinks, fluxes, trends, and interactions involving the mobile biogeochemical constituents within the earth system, including human activities, with a focus on carbon, nitrogen, sulfur, oxygen, phosphorus, and the halogens.

• Ecological systems and dynamics. The investigation of the responses of ecological systems, both marine and terrestrial, to changes in global and regional environmental conditions and of the influence of biological communities on the atmospheric, terrestrial, oceanic, and climate systems.

• Earth system history. The uncovering and interpretation of the natural records of past environmental change that are contained in terrestrial and

INTRODUCTION

marine sediments, soils, glaciers and permafrost, tree rings, rocks, geomorphic features, and other direct or proxy documentation of past global conditions.

- Human interactions. The study of (1) the social factors that influence the global environment, including population growth, industrialization, agricultural practices, and other land usages, and (2) the human activities that are affected by regional aspects of global change.
- Solid earth processes. The study of geological processes (e.g., volcanic eruptions and erosion) that affect the global environment, especially those processes that take place at the interfaces between the earth's surface and the atmosphere, hydrosphere, cryosphere, and biosphere.
- Solar influences. The investigation of how changes in the near-space and the upper atmosphere that are induced by variability in solar output influence the earth's environment.

The U.S. President's budget for fiscal year 1991 proposed a budget of over $1 billion for this program, with the participation of seven federal agencies.

INTERNATIONAL PROGRAMS: IGBP AND WCRP

Internationally, scientific research on global change is being undertaken principally under the auspices of two complementary scientific programs—the IGBP, of the International Council of Scientific Unions (ICSU), and the WCRP, which is jointly carried out under the auspices of the World Meteorological Organization (WMO) and ICSU.

The IGBP was formally adopted by ICSU in 1986 with the objective to (ICSU, 1986)

> describe and understand the interactive physical, chemical, and biological processes that regulate the total earth system, the unique environment that it provides for life, the changes that are occurring in this system, and the manner in which they are influenced by human activities.

Priority in the IGBP falls on

> those areas of each of the fields involved that deal with the key interactions and significant change on time scales of decades to centuries, that most affect the biosphere, that are most susceptible to human perturbation, and that most likely lead to predictive capability.

Since 1986, ICSU's Special Committee for the IGBP has developed several operational components of the program. These core projects include the International Global Atmospheric Chemistry (IGAC) program, addressing the important interactions between the terrestrial and marine biospheres and the atmosphere; the Joint Global Ocean Flux Study (JGOFS), addressing the fluxes of carbon and associated biogenic elements in the ocean and the exchanges with the atmosphere, the sea floor, and the continental boundaries; Past Global Changes (PAGES), aiming to improve understanding of the history of the earth system over the past 2,000 years and the dynamics that caused glacial-interglacial variations in the late Quaternary epoch; the biospheric aspects of the hydrological cycle; terrestrial systems and global change; and data and information needs. Other core projects focusing on such issues as coastal zones and earth system models are currently being developed (IGBP, in preparation).

The World Climate Program was established by WMO in 1979, on the basis of extended studies by expert panels of its executive committee and the deliberations of the First World Climate Conference. The program has four major components, dealing with data, applications, the study of impacts, and research. The last of these, the WCRP, builds directly upon the scientific and institutional framework of the highly successful Global Atmospheric Research Program (GARP). Like GARP, it is conducted as a joint enterprise of WMO and the nongovernmental ICSU.

The major objectives of the WCRP are to determine

- to what extent climate can be predicted and
- the extent of man's influence on climate.

To this end, the WCRP initiates studies of regional and global climate, climate variability, and mechanisms; assesses significant trends; develops physical-mathematical models; and investigates the sensitivity of climate to natural and human stimuli. Planning currently considers three principal streams of activity dealing with various time scales: monthly/seasonal; seasonal/interannual; and long-term climate sensitivity and change over decades to centuries.

Within this framework, a number of projects of varying degrees of relevance to the global change effort are being planned or implemented. Among these are the World Ocean Circulation Experiment (WOCE), the Tropical Ocean and Global Atmosphere (TOGA) program, the Global Precipitation Climatology Project, the International Satellite Cloud Climatology Project, the International Satellite Land Surface Climatology Project, and the Global Energy and Water Cycle Experiment (GEWEX).

OBJECTIVES AND ORGANIZATION OF THIS REPORT

This report recommends a number of initiatives for achieving the goals and objectives of the U.S. Global Change Research Program. Specifically, the report elaborates on the scientific needs for developing integrated earth system models, conducting focused studies to improve understanding of global change, and documenting global change. The recommended focused studies, previously identified by the Committee on Global Change (NRC, 1988), constitute the initial priorities for the U.S. contributions to the IGBP.

The report is not intended to provide a comprehensive view of all the scientific priorities for the USGCRP. Rather, it represents a contribution to the full suite of scientific planning documents for global change research, including reports of the interagency Committee on Earth Sciences (CES, 1989a,b, 1990), the World Climate Research Program (WCRP, 1990), and the International Geosphere-Biosphere Program (1986, 1988, in preparation).

In chapter 2 of this report, the committee addresses the scientific needs that must be met to develop integrated conceptual and predictive models of the earth system. Chapters 3 through 7 address the objective of the USGCRP to develop focused studies to improve understanding of earth system processes, and recommend specific research initiatives in five areas that should be given priority attention. These initiatives focus on the interdisciplinary research required to improve understanding of the interactions between the oceans, atmosphere, land, and human activities, and assume effective support for more disciplinary-oriented research to further understanding for relevant processes. In chapter 8, the committee recommends specific efforts to establish an integrated, comprehensive long-term program of documenting the earth system on a global scale.

The interdisciplinary initiatives recommended for priority attention and discussed in chapters 3 through 7 are as follows:

- earth system history and modeling to document changes in atmospheric composition, climate, and human activities to improve and validate models of global change.
- human interactions with the geosphere-biosphere to analyze changes in human land use, energy use, and industrial processes that drive changes in the earth system.
- water-energy-vegetation interactions to develop global models of the response of terrestrial ecosystems to changes in climate, land, and water use and to determine the reciprocal effects of such changes in terrestrial ecosystems on the climate system on regional and global scales.
- fluxes of materials from terrestrial ecosystems to improve understanding of the processes most important for determining fluxes of radiatively active gases between the land and the atmosphere, in order to predict how changes in climate and land use alter gas emissions, and to improve understanding of the effects of land use changes on nutrient transfer to river, estuarine, and ocean systems.
- biogeochemical dynamics in the ocean interactions with climate in order to predict effects of climatic change on oceanic biogeochemical cycles

and the interactions of such cycles with climate via the release and absorption of radiatively active gases.

Each of these initiatives includes process studies, modeling efforts, and data collection relevant to the particular issues under study.

Figure 1.1 shows the relationship between the initiatives recommended in this report, the elements of the USGCRP, and projects of the IGBP and

ICSU International Geosphere-Biosphere Program Coordinating Panels, Working Groups, & Scientific Steering Committees	Committee on Global Change Recommended Initial Foci	Elements of the U.S. Global Change Research Program	World Climate Research Program Streams
Global Geosphere-Biosphere Modeling	Integrated Earth System Models		
Data Management & Information Systems	Earth System Measurements	Earth System Measurements & Data	Stream I: Monthly/Seasonal Time Scale
Global Changes of the Past	Earth System History & Modeling	Earth System History	Stream II: Seasonal/Interannual (Tropical Ocean-Global Atmosphere)
Terrestrial Biosphere-Atmosphere Chemistry Interactions	Terrestrial Trace Gases & Nutrient Fluxes	Biogeochemical Dynamics	
Marine Biosphere-Atmosphere Interactions	Biogeochemical Dynamics in the Ocean	Ecological Systems & Dynamics	
Biospheric Aspects of the Hydrological Cycle	Water-Energy-Vegetation Interactions	Climate and Hydrological Systems	Stream III: Decades to Centuries (World Ocean Circulation Experiment, Global Energy Water Experiment)
Effects of Climatic Change on Terrestrial Ecosystems	Human Interactions	Human Interactions	
Geosphere-Biosphere Observatories		Solid Earth Processes	
		Solar Influences	

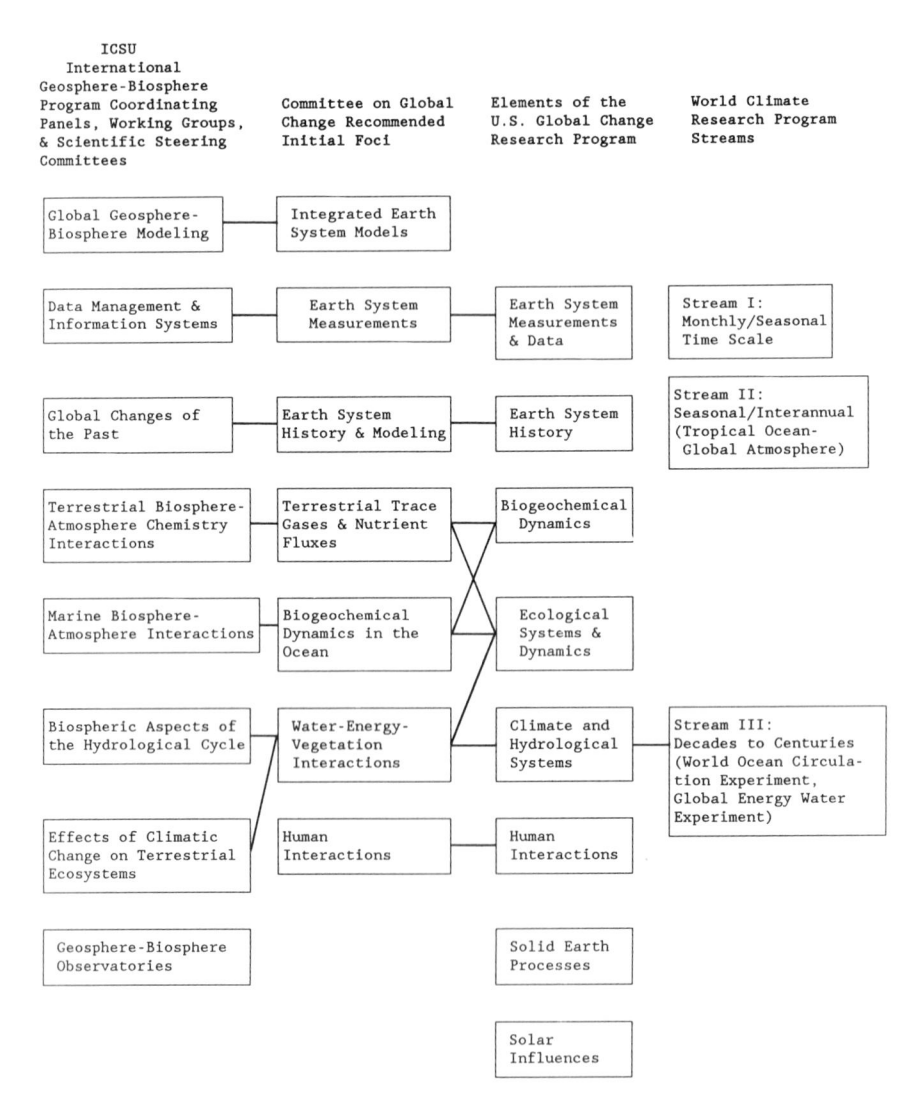

FIGURE 1.1 Relationships of scientific themes for research programs on global change.

WCRP. The initiatives recommended in this report should be viewed in the context of the full suite of scientific investigations required by the USGCRP. Thus the recommendations presume the effective support of global change investigations across a broad spectrum including physical climate and hydrological systems, solid earth processes, and solar influences. As scientific understanding evolves, new priorities will emerge and be refined and incorporated into specific projects.

REFERENCES

Committee on Earth Sciences (CES). 1989a. Our Changing Planet: A U.S. Strategy for Global Change Research. Federal Coordinating Council for Science, Engineering, and Technology. Office of Science and Technology Policy, Washington, D.C.

Committee on Earth Sciences (CES). 1989b. Our Changing Planet: The FY 1990 Research Plan. Federal Coordinating Council for Science, Engineering, and Technology. Office of Science and Technology Policy, Washington, D.C.

Committee on Earth Sciences (CES). 1990. Our Changing Planet: The FY 1991 U.S. Global Change Research Program. Federal Coordinating Council for Science, Engineering, and Technology. Office of Science and Technology Policy, Washington, D.C.

International Council of Scientific Unions (ICSU). 1986. The International Geosphere-Biosphere Program: A Study of Global Change. Report No. 1. Final report of the Ad Hoc Planning Group. ICSU Twenty-first General Assembly, Sept. 14-19, 1986, Bern, Switzerland.

International Geosphere-Biosphere Program (IGBP). 1988. The International Geosphere-Biosphere Program: A Study of Global Change. A Plan for Action. A report prepared by the Special Committee for the IGBP for discussion at the First Meeting of the Scientific Advisory Council for the IGBP, Oct. 1988, Stockholm, Sweden. Rep. 4. IGBP Secretariat, Stockholm.

International Geosphere-Biosphere Program (IGBP). 1990. The Initial Core Projects. Report to the Second Scientific Advisory Council for the IGBP. In preparation.

National Research Council (NRC). 1988. Toward an Understanding of Global Change: Initial Priorities for U.S. Contributions to the International Geosphere-Biosphere Program. National Academy Press, Washington, D.C.

World Climate Research Program. 1990. Global Climate Change: A Scientific Review. Presented by the World Climate Research Program. World Meteorological Organization and International Council of Scientific Unions.

2
Integrated Modeling of the Earth System

OVERVIEW

The possibility of major changes in the global environment presents the scientific research community with a difficult task: to devise ways of analyzing the causes of and projecting the course of these shifts as they are occurring. Purely observational approaches are inadequate for providing the needed predictive or anticipatory information because response times of many terrestrial ecosystems are slow and there is a great deal of variability from place to place. Furthermore, many important processes cannot be measured directly over large areas, such as those processes that occur in soils. We need models to express our understanding of the complex subsystems of the earth and how they interact with and respond to and control changes in the physical-climate and biogeochemical systems.

By the year 2000, a fully coupled, dynamical model of the earth system (Figure 2.1) could be a reality. Such models would significantly improve capabilities for projecting changes in the earth system on a decadal time scale. The focus of this chapter is on the efforts required to achieve this goal. For instance, it is necessary to begin now to develop models that are more completely coupled—albeit still partial—than those that are currently available. Even though these prototypes may themselves not be successful,

---

This chapter was prepared by the working groups on Integrated Earth System Models established under the Committee on Global Change. Members of the group on Terrestrial-Atmosphere Modeling were Berrien Moore III, University of New Hampshire, Chair; John Aber, University of New Hampshire; Guy Brasseur, National Center for Atmospheric Research; Robert Dickinson, National Center for Atmospheric Research; William Emanuel, Oak Ridge National Laboratory; Jerry Melillo,

INTEGRATED MODELING OF THE EARTH SYSTEM 17

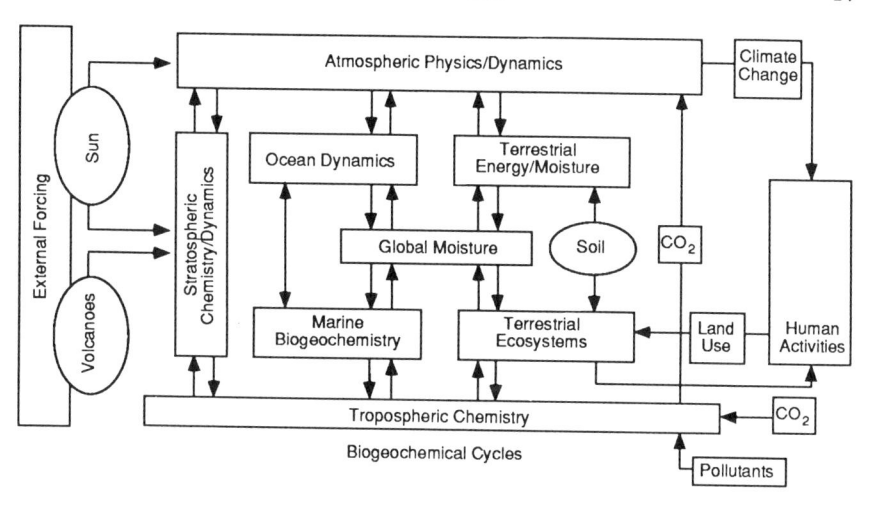

FIGURE 2.1 Status of earth system science in the year 2000 (ESSC, 1988).

they will teach us what is needed to realize our goal of a fully coupled, dynamical earth system model with a multidecadal scale of analysis. However, it should not be overlooked that much of the real science is in the simple models and empirical observations that guide our understanding and give us a framework for interpreting (and creating) the more complex models that evolve later. The early linking of complex models and the subsequent addition of existing approaches should be balanced by efforts to create new, insightful simple models. Such insights provide the basis for qualitative improvements in model structures.

It should be recognized at the outset that the multidecadal temporal scale places important constraints and demands upon the character of earth system models (Bolin et al., 1986; ESSC, 1988; NRC, 1988). For instance, the

---

Marine Biological Laboratory; David Schimel, Colorado State University; Piers Sellers, University of Maryland; and Herman Shugart, University of Virginia. Members of the group on Ocean-Atmosphere Modeling were Berrien Moore III, University of New Hampshire, Chair; Mark Abbott, Oregon State University; Curt Covey, Lawrence Livermore National Laboratory; Nick Graham, Scripps Institution of Oceanography; Dale Haidvogel, Johns Hopkins University; Eileen Hoffman, Old Dominion University; Christopher Mooers, University of New Hampshire; James O'Brien, Florida State University; Albert Semtner, Naval Postgraduate School; and Leonard Walstad, Oregon State University. Members of the group on Atmospheric Physics-Atmospheric Chemistry were Berrien Moore III, University of New Hampshire, Chair; Guy Brasseur and Robert Dickinson, National Center for Atmospheric Research; Bill Gross, NASA Langley Research Center; and Chris Morris, University of New Hampshire.

temporal scale demands inclusion of the biosphere and coupling across critical interfaces: terrestrial ecosystems and the atmosphere, the chemistry of the atmosphere and the physics of the atmosphere, and the oceans and the atmosphere. Advances at these interfaces are essential for progress.

Differences in characteristic rates of change and fundamental processes of different components of the system will impose subsystem-specific demands and requirements on component models (Rosswall et al., 1988). Ecological systems will most likely rest upon functional groups rather than species; understanding biogeochemical fluxes will require process-level models, but initial implementation at global scales will certainly require extensive parameterization. Similarly, the nonlinear chaotic dynamics of the fluid subsystems—the oceans and atmosphere—will continue to require a careful, step-by-step buildup in complexity; the simplistic thinking that must go into all initial modeling advances will tend to be eventually superseded by computationally intensive three-dimensional approaches. This is, in fact, occurring in many of the geophysical and biological-biogeochemical sciences.

The most complex models to date are the atmospheric and oceanic general circulation models (GCMs). These have structures largely determined by the need to solve the Navier-Stokes fluid equations, but they are rich in other physical processes as well. The atmospheric models and their climate role are especially strongly governed by water processes; however, it is precisely these aspects, including questions of scale and parameterization, that are among the least satisfactory of the models.

Resolution is a problem in that the spatial scales of many of the important atmospheric water structures are poorly resolved by existing models. For example, many of the cloud systems that are most important for atmospheric radiation have vertical scales of less than the thickness of the layers in most existing GCMs. The horizontal structure of precipitating systems suffers not only from inadequate resolution but also from severe difficulties with the currently available numerical schemes that were designed primarily for effectiveness (minimal computational demands) in treating the model hydrodynamics. One obvious defect of these schemes is the tendency of truncated spectral series to give negative mixing ratios for water in high latitudes, a consequence of the failure of the series to represent properly the fields in going from relatively large mixing ratios to relatively small ones. The same difficulty can be encountered for any model tracer. For example, it was difficult to get models to treat global smoke fields properly in nuclear winter computations. The hope is that the new semi-Lagrangian schemes will cure these numerical difficulties.

Another question in the treatment of water vapor in various atmospheric GCMs is whether vertical transport in the models resembles the process in nature, again because much of the real vertical transport occurs on scales that are small in comparison with that of the model. The subgrid-scale moist convection parameterizations in the models are still fairly crude and have not improved much in the last decade, although considerable effort is now going into them (Anthes, 1983).

Adding the important chemical constituents and the reactions to an atmospheric GCM causes the issues of scale and computational challenges to become daunting. Many of the important chemical reactions are concentration dependent and hence grid-scale dependent, and important processes often occur in the boundary layer, which generally is not well enough resolved. Further, the addition of atmospheric chemistry to a GCM places greater demands upon the terrestrial and oceanic boundary conditions and dynamic simulations (Lenschow and Hicks, 1989; NRC, 1984; Schimel et al., 1989).

In considering coupling atmospheric GCMs to terrestrial models, where the coupling transfers not only energy and water but also important gases, such as carbon monoxide, methane, and carbon dioxide for the carbon cycle, temporal- and spatial-scale issues again emerge. The macrobalance of terrestrial carbon stocks, which determine the net flux of carbon dioxide, are difficult to derive by integrating across the short time scales at which energy, water, and carbon dioxide and oxygen are actually exchanged because of the high degree of variability that these processes exhibit. Longer time step integrations have generally been more successful. On the other hand, the flux of methane and other short-lived species cannot be treated by simple mass balance and crudely time-averaged responses. Ecological changes, such as successional sequences of tree species, are not well treated on time steps that are appropriate for considering photon input and water exchange or even trace gas fluxes and require some intermediate parameterization or model.

The relatively simple coupling issue of land hydrology and atmosphere remains elusive, and yet it is quite important. The exchange of many reduced gases (e.g., methane) depends on soil moisture conditions, and energy fluxes are influenced by water balances. Modeling sensitivity studies have shown that if evapotranspiration were turned off over continental-scale areas, summer precipitation would be severely reduced and temperatures would be as much as 10 K higher than with normal fluxes. They also show that over tall vegetation the integrated resistance to transpiration implied by the stomata will have a major effect on Bowen ratios over the diurnal cycle. Since the rates of sensible heat exchange over the diurnal cycle determine the height reached by the planetary boundary layer as well as diurnal variations of precipitation in tropical and summer conditions, it is evident that it is important to include the role of vegetation in simulations of the hydrological cycle. Better field data are helping to establish the parameters needed for linking plant physiology to surface evapotranspiration, but considerable further effort is needed before the appropriate submodels can be applied with confidence over a wide range of vegetation cover (e.g., Dickinson, 1984; Eagleson, 1986; Sellers et al., 1986).

The coupling between the ocean and the atmosphere is central to the question of climate change. Atmospheric GCMs with prescribed oceans, long the mainstay of three-dimensional climate modeling, are inherently incapable of simulating the actual time-evolving response of the climate system to increasing greenhouse gases because this response involves heat uptake by the oceans. This is particularly clear when one realizes that the heat capacity of the atmosphere is roughly equivalent to that of the upper 3 m of the ocean. While it is true that the ocean may, partially, act in a passive manner, studies of the El Niño/Southern Oscillation (ENSO) show that the ocean-atmosphere system responds in a coupled fashion on interannual time scales, and paleo-oceanographic investigations suggest that aspects of longer-term climate change are associated with changes in the ocean's thermohaline circulation. The capability to predict these changes in circulation and heat exchange is necessary to describe the future evolution of global climate (e.g., Bryan et al., 1982; Cess and Goldenberg, 1981; IPCC, 1990; Sarmiento et al., 1988).

Fortunately, exciting and encouraging progress is being made in coupling key aspects of the major subsystems. Results from linking atmospheric and oceanic GCMs have already been reported in the literature and have shown significantly different behavior from that of simulations in uncoupled modes. Similarly, interactive simulations between atmosphere and land vegetation have been reported, and these have also exhibited new dynamical characteristics. The inclusion of biology in oceanic GCMs has begun, although the models are still simplistic and do not yet include climatic feedback in a coupled system. Representations of terrestrial biology are also preliminary and again without critical biogeochemical feedbacks. Finally, progress is being made toward model structures and data sets that will allow implementation of atmospheric-oceanic-terrestrial models that include key biological-biogeochemical feedbacks.

For the near term, developments in modeling the earth system should continue to focus on linking previously unlinked components, adding specific subsystems to existing models (e.g., coupling oceanic and atmospheric GCMs or adding a marine biospheric model to an oceanic GCM), or improving existing linked treatments. In this spirit, the committee has arranged the following discussion around three interface models:

1. Atmosphere-terrestrial subsystem.
2. Physical-chemical interactions in the atmosphere.
3. Atmosphere-ocean subsystem including interactions with the biosphere.

These three subsystems obviously overlap and do not include all interfaces. Further consideration is required on the issue of the role of the cryosphere and its coupling on multidecadal time scales (see OIES, 1989).

In the following sections the committee presents a brief general discussion of the current status of models at these three interfaces, including for each a focused report on recommended initiatives and themes. The two final sections deal with the cross-cutting issues of model tests and infrastructure. In order to provide perspective on the remainder of the chapter, the following considerations for each of the three interface models are provided:

For models that couple the terrestrial ecosystems and the atmosphere:

• The coupling must address questions such as how will a changing climate affect terrestrial carbon dioxide uptake and storage; how will evapotranspiration change; how will the distribution of vegetation and its seasonal pattern change; what are the effects on climate of changing patterns of vegetation, including large-scale deforestation; and what is the effect of changing chemical conditions on terrestrial vegetation and trace gas exchange?

• The primary research issue in understanding the role of terrestrial ecosystems in global change is that of analyzing how processes with vastly differing rates of change, from photosynthesis to community change, are coupled to each other and to the atmosphere.

• Modeling these interactions requires coupling successional models to biogeochemical models to physiological models. Of these, only the physiological models can currently describe the exchange of water and energy between the vegetation and the atmosphere at fine time scales.

• Terrestrial models should focus on linked models addressing plant community change, biogeochemistry, and physiology and biophysics. Models of the physics of the atmosphere couple directly to terrestrial physiology models; biogeochemical models serve as a bridge between physiology and community change as well as coupling to the chemistry of the atmosphere.

• The coupling must address how changes in the global environment, including the effects of land use and chemical stress, affect terrestrial ecosystems and how ecosystem changes affect the global system.

• Formidable problems of scale and parameterization are raised in three- and four-dimensional simulations of biology and atmospheric chemistry because of nonlinear concentration-dependent phenomena.

For models that couple physics and chemistry in the atmosphere:

• The coupling must address questions such as what is the spatial-temporal distribution of carbon monoxide, methane, and tropospheric ozone and how might it change; what is the effect of changing climatic or chemical conditions on the aerosol-initiated stratospheric ozone depletion in the Arctic and the Antarctic; how might the exchange of water vapor between the troposphere and the stratosphere change in a changing climate; and what is the vertical transport of trace species by cloud convection and how might it change?

- Progress in the modeling of the coupled chemical-physical atmospheric system requires a better knowledge of surface sources of trace gases and their dependence on climatic conditions; chemical processes and reaction channels, both in the gas and in the aqueous phase, and their dependence on atmospheric conditions; and transport processes by advection and convection, including the development of high-resolution transport models coupled to atmospheric GCMs with detailed representation of physical processes including cloud formation and associated transport, boundary layer transport, and troposphere-stratosphere exchange. This progress is dependent on the acquisition of global data sets for validation of these treatments.
- Future progress will be dependent both on available computational resources and on progress in developing our understanding of fundamental physical and chemical processes and the nature of their coupling.

For models that couple the ocean and the atmosphere:

- The coupling must address questions such as how will changing climate affect oceanic carbon dioxide uptake and storage; how will oceanic heat storage and transport change; how will the amount and distribution of primary production change; how will the marine hydrological cycle change; and how will a changing ocean affect a changing climate?
- Critical issues include widely differing temporal and spatial scales, inclusion of biological and biogeochemical dynamics, and sparse data. Particularly important and difficult tasks are the scaling of the biological-biogeochemical components from local-regional domains to basin-global domains, formation of the upper mixed-layer physics, and inclusion of possible biological feedbacks on mixed-layer dynamics.
- Progress in the development of coupled oceanic-atmospheric models including biological-biogeochemical dynamics is limited, in part, by an inadequate theoretical or observational understanding of certain key processes and a corresponding and continuing uncertainty as to how best to incorporate or parameterize them in oceanic GCMs.
- The set of field programs (JGOFS, WOCE, the Coupled Oceans Atmosphere Research Experiment (COARE) organized under TOGA, and the Global Ocean Ecosystem Dynamics (GLOBEC)) required to acquire the data needed to advance our knowledge of fundamental oceanic processes is already well defined. These programs also offer valuable opportunities for simultaneous observational efforts, and these should be encouraged.
- The development of fully coupled models should be encouraged along two parallel paths: the first devoted to developing basin- and global-scale models with increasing levels of coupling, and the second leading to a series of regional fine-scale models that could provide boundary conditions and parameterization tests for the larger-scale models.

Several overarching issues exist regarding approaches to and testing of models and the infrastructure necessary for their development:

- Validation is extremely difficult; models should be subjected to naturally occurring perturbation tests that exercise the coupling. In addition, large-scale phenomena offer a valuable opportunity for focusing model developments and testing model dynamics. Studies of these large-scale processes will serve not only as diagnostic tests but also as prognostic tools.
- It is urgent that testing of models and model combinations begin as soon as possible. Experiments with global models will initially use simple representations, but the lessons learned and data bases developed will be critical to future improvements. Prototype global experiments will be especially important to exploring feedbacks between the production of long-lived trace gas species and climate.
- Two important themes are important in early testing of partial earth system models: the global carbon cycle (carbon dioxide, methane, and carbon monoxide) and the transient response to a changing greenhouse forcing. The former exercises the chemistry and biology, whereas the latter stresses the physics and biology. The obvious next step is coupling these two themes.
- The importance of experience gained through prototype modeling experiments, including failure, should not be underestimated. Careful analysis of failures can provide valuable information.
- Earth system modeling should serve as a focus and catalyst for interdisciplinary science. No one institution or group of investigators has more than a fraction of the interdisciplinary talent necessary for the development of an earth system model focused on multidecadal time scales. Thus several teams and talented individuals should be supported, who with some coordination could help perform the incremental steps toward the integrated earth system model. Some of these groups may act primarily as synthesizers, their principal interest being in linking component pieces, while in other groups the interest would be in component development.
- Also needed in an overall modeling strategy are centralized facilities and associated staff to serve the common needs of the various teams and individuals and focus on issues of synthesis, continuity, documentation, and extensive numerical experiments.

ATMOSPHERE-TERRESTRIAL SUBSYSTEM

The primary research issue for coupling atmosphere-terrestrial models is understanding how processes with vastly differing rates of change, from photosynthesis to community change, are coupled. Representing this coupling in models is the central challenge to modeling the terrestrial biosphere as part of the earth system (e.g., Allen and Wyleto, 1984; Huston et al., 1988; King et al., 1990; Moore et al., 1989b; Smith et al., 1989).

Terrestrial ecosystems participate in climate and in the biogeochemical cycles on several temporal scales. The metabolic processes that are responsible for plant growth and maintenance, and the microbial turnover associated with dead organic matter decomposition, move carbon and water through rapid as well as intermediate time scale circuits in plants and soil. Moreover, this cycle includes key controls over biogenic trace gas production. Some of the carbon fixed by photosynthesis is incorporated into plant tissue and is delayed from returning to the atmosphere until it is oxidized by decomposition or fire. This slower carbon loop through the terrestrial component of the carbon cycle, which is matched by cycles of nutrients required by plants and decomposers, affects the increasing trend in atmospheric carbon dioxide concentration and imposes a seasonal cycle on that trend (Figure 2.2). The structure of terrestrial ecosystems, which responds on even longer time scales, is the integrated response to the intermediate time scale carbon machinery. The loop is closed back to the climate system since it is the structure of ecosystems, including species composition, that sets the terrestrial boundary condition in the climate system from the standpoint of surface roughness, albedo, and, to a great extent, latent heat exchange.

These separate temporal scales contain explicit feedback loops that may modify the system dynamics. Consider again the coupling of long-term climatic change with vegetation change. Climatic change will drive vegetation dynamics, but as the vegetation changes in amount or structure, this

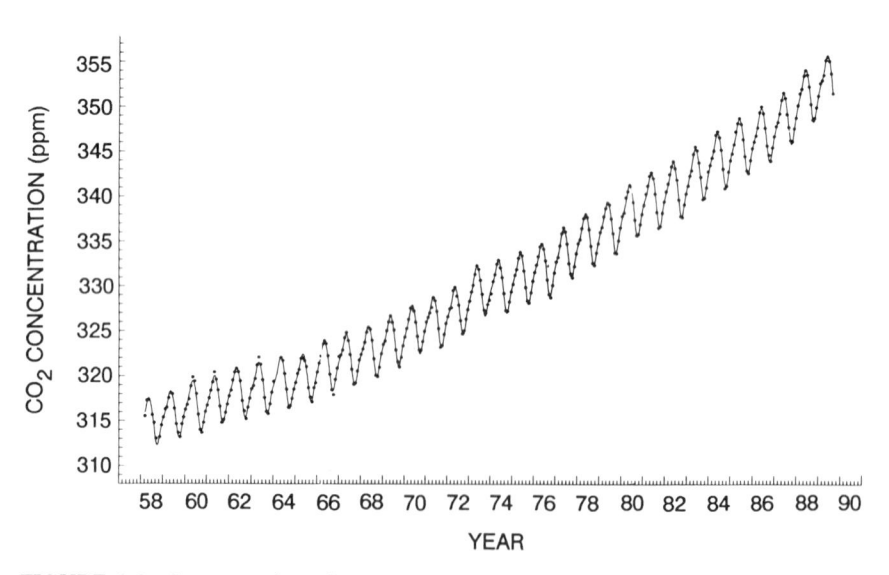

FIGURE 2.2 Concentration of atmospheric carbon dioxide in parts per million of dry air (ppm) versus time for the years 1958 to 1989 at Mauna Loa Observatory, Hawaii. The dots indicate monthly average concentration. (From C.D. Keeling et al. (1989). Copyright © 1989 by the American Geophysical Union.)

will feed back to the atmosphere through changing water, energy, and gas exchange. Biogeochemical cycling will also change, altering the exchange of trace gas species. The long-term change in climate, driven by chemical forcing functions (carbon dioxide and methane) will drive long-term ecosystem change. Modeling these interactions requires coupling successional models to biogeochemical models to physiological models that describe the exchange of water and energy between the vegetation and the atmosphere at fine time scales. There does not appear to be any obvious way to allow direct reciprocal coupling of GCM-type models of the atmosphere, which inherently run with fine time constants, to ecosystem or successional models, which have coarse temporal resolution, without the interposition of a physiological model. This is equally true for biogeochemical models of the exchange of carbon dioxide and trace species. This cross-time-scale coupling is important and sets the focus for the modeling strategy.

A Modeling Strategy: Prognosis for Progress

Intuitively, we might develop a global model of terrestrial ecosystem dynamics by combining descriptions of each of the physical, chemical, and biological processes involved in the system. In such a scheme, longer-term vegetation changes would be derived by integrating the responses of rapidly responding parts of the model. But we cannot simply integrate models that describe the rapid processes of carbon dioxide diffusion, photosynthesis, fluid transport, respiration, and transpiration in cells and leaves in order to estimate productivity of whole plants, let alone entire ecosystems. The nature of the spatial averaging implied in the selection of parameters and processes to consider is difficult because of nonlinearities, which means that the choice of scale influences the calculation of averages (see Rosswall et al., 1988).

To progress in the development of terrestrial ecosystem models, we choose processes to treat in different models based on the phenomenological scales involved. As is common in physical models, terms in fundamental equations can be included or ignored depending on the temporal and spatial scales of interest (e.g., ignoring gravitational effects in quantum physics and including Coriolis effects in large-scale fluid motion). Careful organization of a suite of models, each describing processes that operate at different rates, is crucial to the practical development of terrestrial ecosystem models for use in earth system models of global change.

Based on current model structures, atmosphere-biosphere interactions can be captured with simulations operating with three characteristic time constants (Figure 2.3). The first level represents rapid (seconds to days) biophysical interactions between the climate and the biosphere (Figures 2.4a and b). The dynamics at this level result from changes in water, radiation,

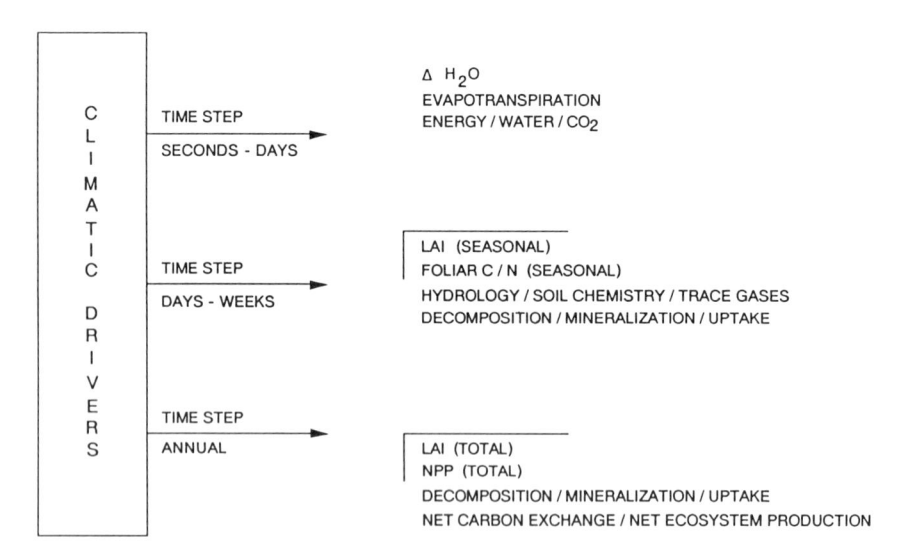

FIGURE 2.3 Three different time steps at which existing models of terrestrial ecosystems use climatic information to modify rates of ecosystem function.

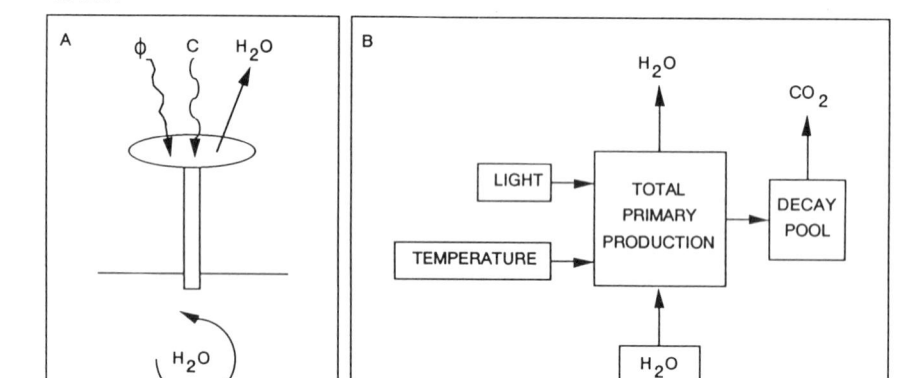

FIGURE 2.4 Two diagrammatic representations of models converting short-time-step environmental data (minutes to hours) into balances of energy, water, and carbon. For these models, ecosystem structure, including leaf display and canopy structure, are fixed. Nutrient fluxes other than emission and consumption of trace gases are not dealt with.

and wind and accompanying physiological responses of organisms. These dynamics occur rapidly in relation to plant growth and nutrient uptake—far more rapidly than species replacement can occur. Simulations at this level are required to provide information to climate models on the exchange of energy, water, and carbon dioxide. Tests of this level of model can be accomplished using experimental methods including leaf cuvettes, micrometeorological observations, and eddy correlation flux measurements (e.g., Farquhar and von Caemmer, 1982; Markar and Mein, 1987; Sellers et al., 1986).

The second level captures important biogeochemical interactions. This level captures weekly to seasonal dynamics of plant phenology, carbon accumulation, and nutrient uptake and allocation (Figures 2.5a and b). Most extant models at this level use integrative measures of climate such as monthly statistics and degree-day sums. Changes in soil solution chemistry and microbial processes can be captured at this level for calculation of trace gas fluxes. Primary outputs from this level of model are carbon and nutrient fluxes, biomass, leaf area index, and canopy height or roughness. This level of model is usually tested in field studies with direct measurements of biomass, canopy attributes, and nutrient pools or fluxes (e.g., Melillo et al., 1982; Parton et al., 1988; Pastor and Post, 1985, 1986; Schimel et al., 1985).

A third level of model represents annual changes in biomass and soil carbon (i.e., net ecosystem productivity and carbon storage) and in ecosystem structure and composition (Figures 2.6a and b). Inputs are calculated indices summarizing the effects of climatic conditions on biomass accumulation and decomposition. Outputs include ecosystem element storage, allocation of carbon and other elements between tissue types, and community structure. This type of model currently represents individual organisms or populations and is difficult to apply at large scales because of computational and data requirements (e.g., Aber et al., 1982; Botkin et al., 1972a,b; King et al., 1990; Shugart, 1984; Shugart and West, 1980). Considerable work will be required to develop large-area implementations. This type of model is validated using a combination of process studies, as described above, but integrated to derive annual fluxes, and comparative studies. The community composition and population dynamics aspects of these models are often validated using paleo-data.

An approach to this coupling is highlighted in Figure 2.7. A level 3 model converts annualized indices of climatic conditions and the current ecosystem state into total leaf area and structure for the next year. Within these total values, the second level calculates the phenology of leaf production and loss, the rate of nutrient mobilization and uptake, and hence the seasonal pattern of ecosystem dynamics. Using these seasonal patterns, the first level converts climatic data into energy and water balances over very short

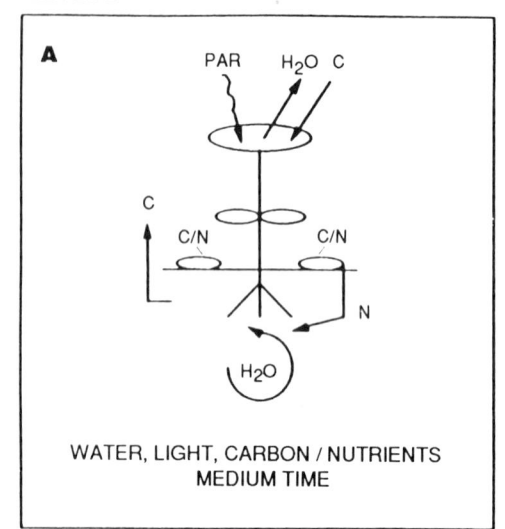

FIGURE 2.5 Two diagrammatic representations of models operating at the intermediate time step (days to weeks). These models capture seasonal phenology and use summary climate data to calculate water, carbon, and nutrient balances at monthly time steps. Energy flux calculations are meaningless at this time step.

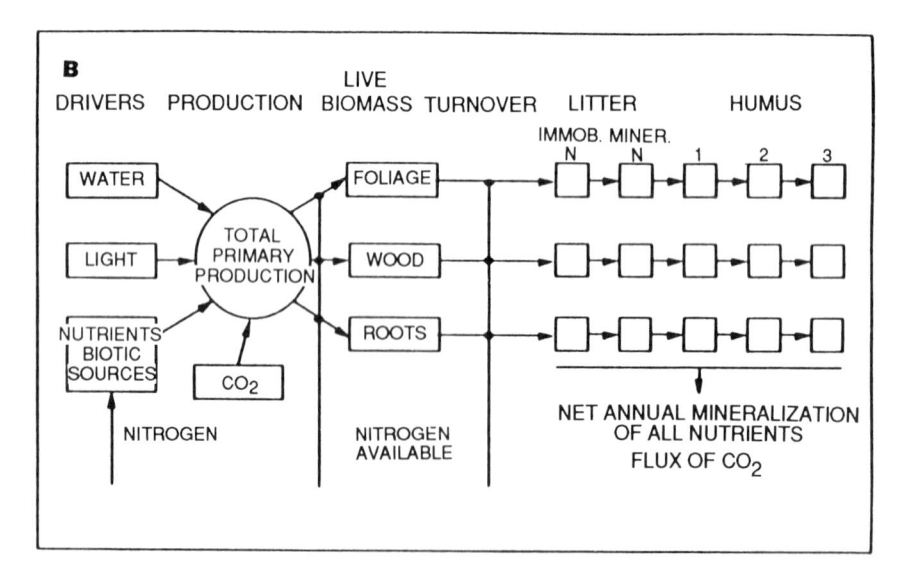

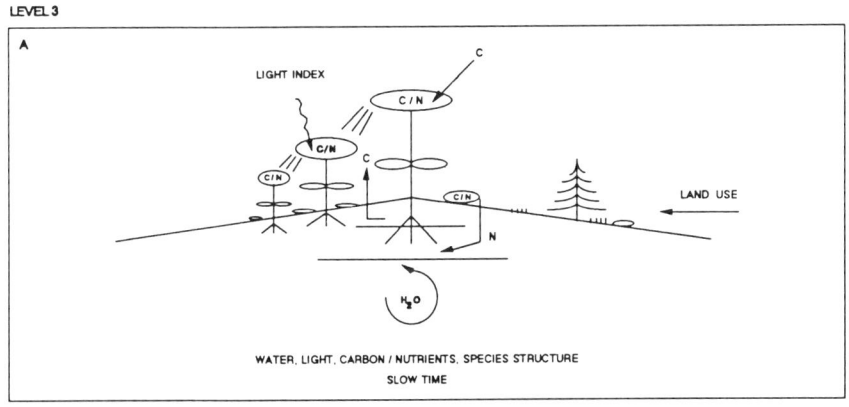

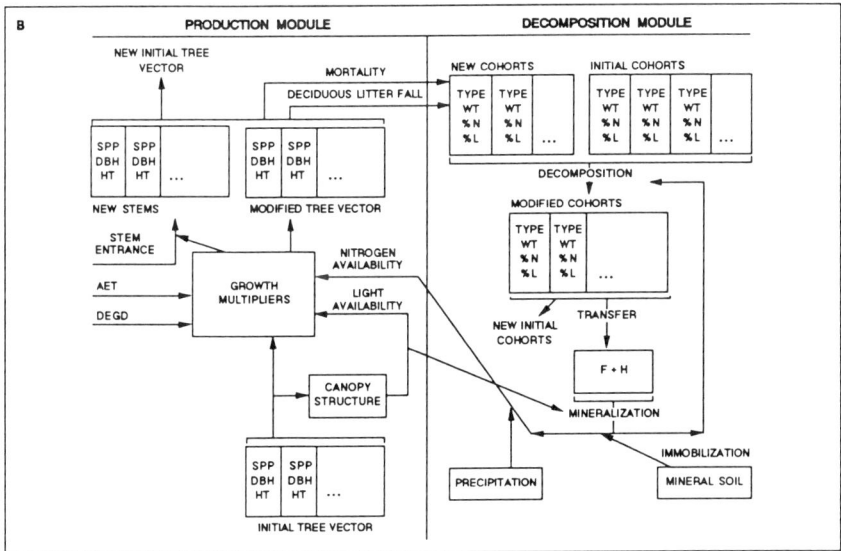

FIGURE 2.6 At the annual time step, models make use of the wealth of data available for growth and mortality of individual stems by species, or growth rate of entire stands of vegetation in relation to annual summations of climatic conditions. Fluxes of carbon and nutrients are calculated. Energy and water balances cannot be dealt with at this time step. NOTE: Acronyms and abbreviations used in this figure are as follows: AET, actual evapotranspiration; DBH, tree diameter; DEGD, growing degree days; F, forest floor; H, organic matter; HT, height; SPP, species; TYPE, litter type (e.g., leaves, wood, root); WT, weight; %N, percent nitrogen; %L, percent lignin.

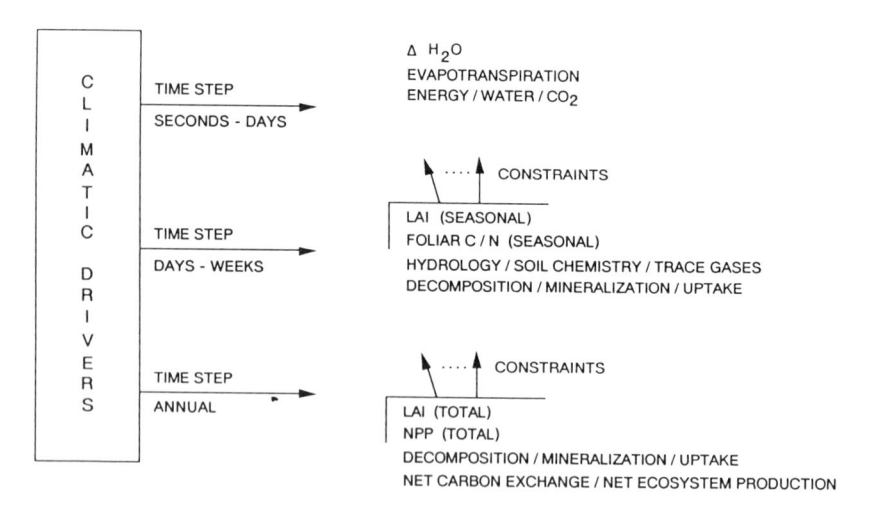

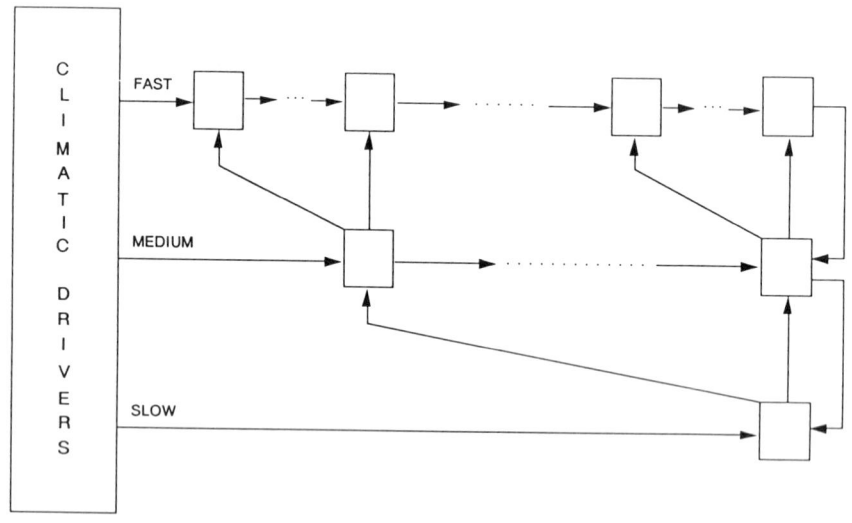

```
                        Δ H₂O
    C                   EVAPOTRANSPIRATION
    L       TIME STEP   ENERGY / WATER / CO₂
    I       ─────────→
    M       SECONDS - DAYS
    A
    T
    I                        ┊····┊ CONSTRAINTS
    C       TIME STEP        LAI (SEASONAL)
            ─────────→       FOLIAR C / N (SEASONAL)
    D       DAYS - WEEKS     HYDROLOGY / SOIL CHEMISTRY / TRACE GASES
    R                        DECOMPOSITION / MINERALIZATION / UPTAKE
    I
    V
    E                        ┊····┊ CONSTRAINTS
    R       TIME STEP        LAI (TOTAL)
    S       ─────────→       NPP (TOTAL)
            ANNUAL           DECOMPOSITION / MINERALIZATION / UPTAKE
                             NET CARBON EXCHANGE / NET ECOSYSTEM PRODUCTION
```

FIGURE 2.7 The use of longer-time-step models to constrain or bound the shorter-time-step efforts. Rather than attempting to run the minute-to-hour models for centuries, it is suggested that the annual-time-step models can be used to estimate changes in structure due to global change for some time into the future. These can then determine the gross ecosystem structure within which the moderate-time-step models can calculate phenological changes and gross water balances. These in turn determine fine-scale ecosystem structure required as inputs for the shortest-time-step models. The shorter-time-step models return indices of physiological stress to the longer-time-step models, which then alter the long-term course of ecosystem development. This approach should provide useful bounds for the shorter-time-step models and reduce computation time considerably.

time steps. Annual feedbacks serve as informational flows carrying back the integrated effect of fine- and medium-scale time steps.

Decomposition calculations can be driven by level 2 vegetation modules and integrated to set nutrient availability in level 3 vegetation calculations. Inorganic soil chemistry routines can operate on almost any time scale, as they tend to be somewhat independent of temperature and linear with time but may be nonlinear with concentrations. Nutrient cycling and soil chemistry modules can run under altered climatic drivers for some time, and then predictions can be made of the consequences of changes in ecosystem state for surface-atmosphere interactions and trace gas fluxes.

This approach is consistent with those outlined in chapters 5 and 6, on water-energy-vegetation interactions and terrestrial trace gas and nutrient fluxes, respectively. Several issues must be resolved to develop and implement such terrestrial modeling schemes in an earth system model of global change:

1. Calculation of indices of climatic effects on biological activity. There are several different ways to summarize the effects of climatic conditions on biological activity. These range from very simple calculations of estimated evapotranspiration, water deficits, and drought indices (Hanks, 1985; Vorosmarty et al., 1989) to physiologically sophisticated and computationally demanding models of water, energy, and carbon balances at the leaf and canopy level (Farquhar and von Caemmer, 1982; Pastor and Post, 1988; Running and Coughlan, 1988; Sellers et al., 1986; Shugart et al., 1986; Solomon, 1986; Solomon et al., 1984). Which of these provides the most accurate depiction of climate-biotic interactions and which can be parameterized most easily from existing or obtainable field data?

2. Spatial scale of species-functional class ecosystem models. Models of forest ecosystem dynamics are of two types: stem-oriented models, which enumerate all individuals within the modeled area, and aggregated models, which deal only with biomass compartments such as foliage, wood, and roots (e.g., see Figures 2.6a and b). Grassland models are generally of the species-aggregated type (e.g., Parton et al., 1988; Schimel et al., 1985). The stem models are valuable for examining gap-phase dynamics within a landscape and can predict changes in species composition explicitly (e.g., Aber et al., 1982; Botkin et al., 1972a,b; Shugart, 1984; Shugart and West, 1980). They are tied to a spatial scale at which canopy gaps occur. The aggregated models are independent of spatial scale and are computationally much simpler but do not capture the successional dynamics or species-specific characteristics of ecosystems (Emanuel et al., 1984; Moore et al., 1989b). At some cost, the stem models can be aggregated spatially through the use of subsampling and geographic information system (GIS) technology to reach GCM spatial scales. Alternatively, the aggregated models can be parameterized to include successional changes.

3. Linking processes across spatial scales. Models need to be organized for use at several levels of geographic detail and with the rapid and slow modules running in concert or separately. At the most detailed geographic level, underlying data are organized on a grid of land cells, and models are solved for each grid cell or with a sampling strategy. Both data and model solutions can be mapped and managed by a GIS. The data requirements and implementation logistics are very demanding at this level. In regional studies, data and model results are tabulated against biome or ecosystem extents, whereas in some applications it is useful to average or lump data and model results to global scale (see Figure 2.7).

An added complexity is that human activities affect a large fraction of the world's terrestrial ecosystems. These disturbances—which range from total management, harvest, and land use change to subtle pollutant impacts—cannot be ignored in any analysis of changes in the role of land systems (e.g., Emanuel et al., 1984; Houghton et al., 1983). Two aspects of the land use perturbation and other human activities must be treated: (1) The rates and distributions of the disturbances per se must be described, perhaps by using a GIS. (2) The effects, particularly the redistribution of carbon and nutrients in the compartments as well as successional patterns, must be described, which is a more difficult issue. This effect of land use change is perhaps best initiated at level 3 and allowed to move upward through the constraint structure (see Figure 2.7).

Research Priorities

As noted above, critical improvements in ecosystem modeling and its linkage to the earth system models will require the development of schemes for integrating processes with very different rates of change. This implies the continued development and validation of physiological, biogeochemical, and successional/population models that are capable of representing the range of processes and communities found in ecosystems worldwide. Experiments with coupling these three levels of models are required, as are tests of these models when run interactively with atmospheric models. The different levels of models have differing data requirements, and these must guide the collection and archiving of data.

In addition to the coupled models, a parallel course needs to be pursued wherein detailed models should be developed that can allow analysis of ecosystem responses to forcing from scenarios of climatic change (i.e., those scenarios developed by exercising the first-generation earth system models). These more detailed models will allow checking of the responses of the simple models included in the earth system models and can be used to develop improved parameterizations for the earth system model. *The key issue here is that the development of process models and parameterizations should proceed together and in a coordinated fashion.*

The above approach leads directly to a strategy for validation (see also the section "Critical Model Tests" below). Process models can often be tested directly in field and laboratory studies and should be so tested before being used to develop parameterizations for use in coupled models. Predictions of the earth system models can also often be tested by comparison to well-mixed, and hence well-sampled, attributes of the atmosphere. Examples of such validation parameters include atmospheric concentrations and latitudinal distributions of atmospheric methane or carbon dioxide. Validation of models at scales between those of field process studies and the globe is often more difficult, requiring intensive and extensive measurements.

It is urgent that tests of models and model combinations begin as soon as possible. Exploration of the behavior of global and regional models with coupled atmosphere and terrestrial systems should begin in parallel. Experiments with global models will initially use simple representations, but the lessons learned and data bases developed will be critical to future improvements. Early global experiments will be especially important to exploring feedbacks between the production of long-lived trace gas species and climate.

Global terrestrial ecosystem models will further our understanding of major phenomena operating within the earth system. Specific extended model studies will address coupled responses of climate and terrestrial ecosystems, perturbation of the global element cycles through the effects of chemical or climatic change on terrestrial ecosystems, and analyses of the effects of land use and terrestrial resource use patterns on the dynamics of climate and element cycling.

Key modeling themes include the following:

• Coupled responses of climate and terrestrial ecosystems. The exchange of heat and moisture between the atmosphere and land systems is an integral part of the climate system that will be altered as terrestrial ecosystems respond to climatic change. Major land cover changes, such as shifts in the distributions of the major biomes, will be accompanied by companion responses in the moisture and heat regimes of the regions involved. Complex transient responses of climate and terrestrial ecosystems toward joint steady state conditions require scrutiny in terms of stability, resource and habitat maintenance, and sensitivity to further perturbation by human activities. The joint responses of these systems will be complicated by broken correlations between environmental variables. For example, while temperature at high latitudes may change, solar intensity and sun angle will not; both variables affect the structure and function of high-latitude vegetation in the northern hemisphere (see chapters 3 and 5).

- Responses of terrestrial element cycles to climatic change. Climatic change, caused by increasing greenhouse gas concentrations, will further perturb carbon cycles in terrestrial ecosystems. Significant additional releases of carbon dioxide and methane into the atmosphere may result, and the cycling of other major elements in addition to carbon will be affected as well. Time constants of major carbon pools on land and the phenology of seasonal variations will be altered. Changes in the geographic patterns in the amplitude and phase of the annual cycle of atmospheric carbon dioxide concentration may be an early indicator of terrestrial ecosystem responses to climatic change. Major model experiments will address these responses of the terrestrial components of the global element cycles to climatic change. The results will contribute to predictions of further greenhouse gas increases and interpretation of satellite data in terms of biogeochemical cycling and may suggest management and resource utilization schemes to minimize the impacts of global change (see chapters 3 and 5).
- Atmospheric chemical forcing on terrestrial element cycling and ecosystem structure and composition. There are already extreme cases of chemical-stress-induced chronic changes. The challenge is to understand and predict the effects of less extreme situations, which may, however, over time exhibit chronic conditions. In addition to considering the effect of chemical stress on terrestrial ecosystems, it will be important to couple the effect of changing climatic conditions as well (see chapter 6).
- Modeling the effect of land use change on terrestrial ecosystems. Future patterns of land and resource use will continue to alter element cycles on land as climatic change occurs. Additional releases from terrestrial pools because of human activities will make a significant contribution to further increases in atmospheric greenhouse gas concentrations. Model studies integrated with satellite data on land cover and its state can monitor the effects of expanding human activity on the terrestrial components of the major element cycles. Finally, experiments using regional ecosystem models and mesoscale climate simulations may begin once we have more sophisticated models.

These experiments will serve as test beds for the early inclusion of sophisticated biology into atmospheric-biospheric models. Use of mesoscale ecosystem-atmospheric simulations will be important to understanding interactions between the biosphere and atmospheric phenomena such as precipitation, boundary layer dynamics, and mesoscale circulation. Efforts in this arena should be coordinated with ongoing field programs such as the Atmosphere Boundary Layer Experiment (ABLE) and the First ISLSCP Field Experiment (FIFE) (see chapters 5 and 6). These intensive field experiments provide the empirical and theoretical framework for this type of modeling; this type of study should form the basis for research in this area. This level of model will also serve to address interactions between the biogenic emissions of short-lived trace gas species and chemical climate. Coordination with the International Global Atmospheric Chemistry (IGAC) program and other ongoing activities could serve as a foundation for this type of activity.

Summary

The changes suggested by current, relatively simple GCMs make it imperative that we seek to couple models of ecological and biogeochemical processes to climate models. Fortunately, the time is now ripe to make this coupling.

Ecological and biogeochemical models consider processes at three characteristic rates of change, and these must be integrated in order to produce appropriate parameterizations for earth system models. Long-time-scale ecological models that emphasize carbon and element storage and population dynamics can be driven by integrations of climate models, but their primary feedback is through the carbon budget and surface roughness. Carbon exchange, net ecosystem production, and trace gas biogeochemistry are best treated at an intermediate time scale and provide feedback to the atmosphere through trace gas exchange and water budgets. Physiological time scale models are required to handle surface exchange of water and energy but must be integrated with other exchanges such as nutrient availability and community type. Examples of these three types of models are now in use; the new challenge is their coupling to atmospheric models. In this process, it is likely that simulations will be both run together and linked only by exchange of parameterizations.

At this stage, simple models (one dimensional, local scale, single process) play a critical role in the development and integration of new science into the dimensional models. It is also likely that development of coupled atmosphere-terrestrial ecosystem models will require some intermediate assembly and testing using regional atmospheric (mesoscale) models. While the multiscale approach appears complex, it provides critical opportunities for validation and testing using techniques that are now available. Although global ecosystem models do not have as much heritage as atmospheric and oceanic circulation models, the science is mature and ready to move into this arena.

Finally, two questions should remain the focus for this atmosphere-terrestrial modeling program:

1. How will global environmental change affect the ecosystems on which mankind depends?

2. How will changing ecosystem structure and function in turn influence the global system through feedbacks to the atmosphere and hydrosphere?

These are both critical issues, and experience suggests that they drive rather different modeling efforts. Both the detailed modeling efforts suggested by the first question and the more heavily parameterized models that are required to run in consort with atmospheric and oceanic models to address the second question are needed and should be pursued in an integrated fashion.

PHYSICAL-CHEMICAL INTERACTIONS IN THE ATMOSPHERE

Models of the physical processes in the atmosphere provide much of our current basis for understanding future climatic change. However, there are still considerable shortcomings in their formulation and implementation, and thus they provide very uncertain projections for the future. Present models have evolved through the combination of two separately developed approaches. On the one hand, they incorporate the contributions of atmospheric dynamics and adiabatic thermodynamics through an approach of computational fluid dynamics that was initially developed in the 1950s to provide an objective numerical approach to weather prediction. It is sometimes forgotten that the initial developments of supercomputers at that time were motivated in large part by the need to solve this problem. The other approach focused primarily on energy balance and was the basis for climate models even in the "back of the envelope" mode.

The thermal/fluid dynamics approach to the weather system has tended to focus on the application of the most efficient and accurate discrete representations of the Eularian, Navier-Stokes, and temperature equations for a compressible atmosphere on a rotating sphere. Meteorological observations are analyzed into initial fields consistent with the model dynamics and then the prognostic variables (e.g., the horizontal winds, temperatures, and surface pressure) are specified from these initial fields and integrated forward in time to generate future weather systems. In the 1960s, versions of these weather prediction models were developed to study the general circulation of the atmosphere, that is, the physical statistics of the weather systems. These models were designed to show how they satisfied global requirements of conservation of momentum and energy. Such model experiments demonstrated that it was necessary to begin to include energy sources and sinks, in particular, exchanges with the surface, moist atmospheric processes (moist convective adjustments and precipitation), and the attendant latent heat release and radiative heat inputs. Until the recent emphasis on the climate problem, the treatment of these physical processes has always been oversimplified, and this is still generally true in current models.

Perhaps the most progress has been made in the treatment of radiative processes, and this has been largely a result of the (initially) independent research stream of radiative-convective energy balance models of the climate system.

The principle of global energy balance is a powerful concept and is the primary basis for our understanding of future global warming (Figure 2.8). In simple terms, the absorption of solar radiation must over a long enough time be balanced by temperature-dependent emission of thermal infrared radiation. For a black body, this emission would be given by the Stefan-Boltzmann law. The point of simple energy balance models is to describe how the global average temperature actually responds to changes in radiative processes, using black body emission as a reference point. In including an atmosphere in these models, it is conventionally assumed that the troposphere is in convective equilibrium, and thus there is only a single degree of freedom for determining temperature from energy balance, whereas the stratosphere and layers above are in radiative equilibrium, with each level adjusting its temperatures according to a local energy balance. The relative simplicity of these models has allowed both the development of conceptual descriptions of the various feedbacks that determine the response of global temperature to changes in radiative forcing and the development of more detailed and accurate radiative models for atmospheric gases and clouds. Hence the initial evaluation of the contributions of various trace gases to global warming has exclusively made use of such models. The effect of changing atmospheric composition, in simplest terms, is estimated by calculating the change of radiative fluxes at the tropopause, assuming the troposphere is unchanged, and then converting this to a temperature change according to our best estimates of climate feedback processes.

In incorporating these concepts into the three-dimensional GCMs, it is imperative that the global energy conservation properties of these models be well understood and interpretable in terms of the simpler climate models. This is not a trivial requirement because the three-dimensional models have many complex energy exchange processes, and so it is easy to introduce spurious energy sources and sinks either through nonconservative numerical procedures or through physical approximations. This might be the case, for example, if a model uses a different treatment of latent heat release for precipitation than it does for surface melting and evapotranspiration. Because of the large number of potential sources of error, it is probably impossible to have a model that conserves energy perfectly, but models should be validated to conserve energy to better than 1 W/m^2 and preferably have errors of less than 0.1 W/m^2. The change of atmospheric radiation from doubling carbon dioxide is about 4 W/m^2, a number that incidently comes from the one-dimensional model approaches referred to above. For an atmospheric model coupled to a surface with ocean temperatures prescribed from observations, the radiative imbalance at the top of the atmosphere should be considerably smaller than this to prevent spurious climatic change

38

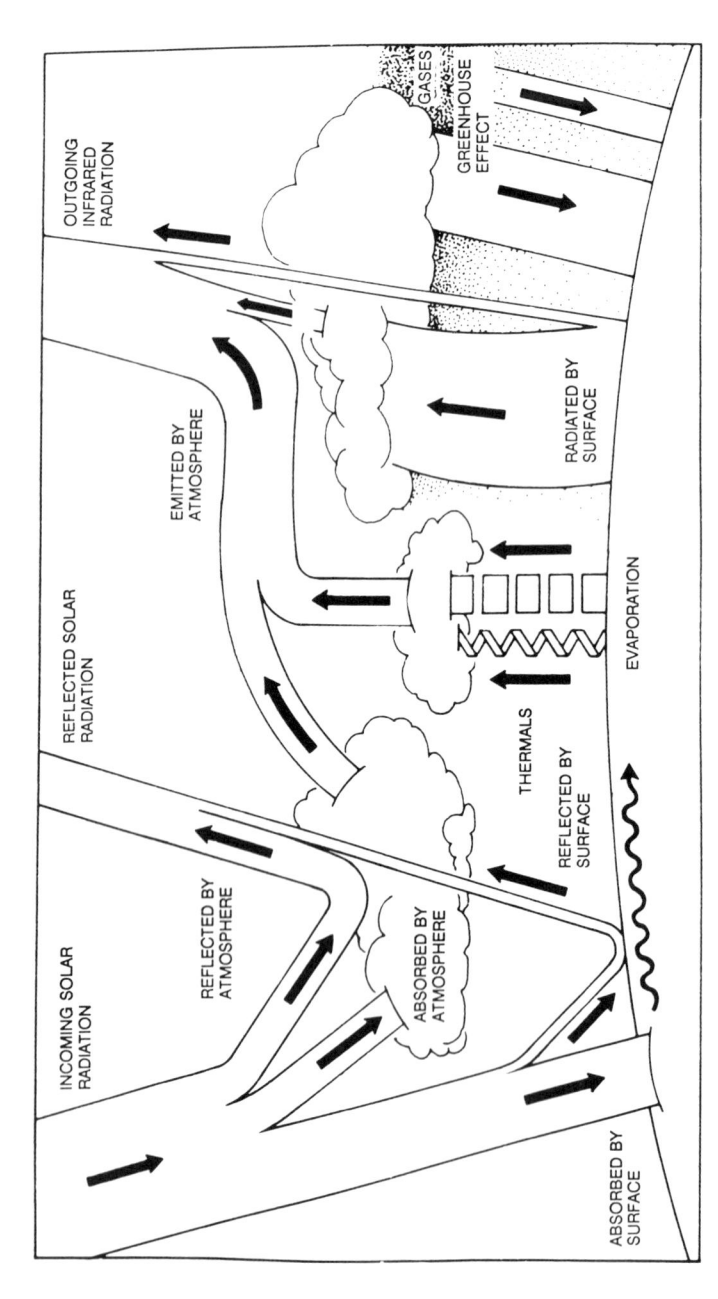

FIGURE 2.8 Schematic illustration of the earth's radiation and energy balances. The greenhouse effect is well established. It arises because the earth's atmosphere tends to trap heat near the surface. Carbon dioxide, water vapor, and other trace greenhouse gases are relatively transparent to the visible and near-infrared wavelengths that carry most of the energy of sunlight, but they absorb more efficiently the longer, infrared (IR) wavelengths emitted by the earth. Hence an increase in the atmospheric concentration of greenhouse gases tends to warm the surface by downward reradiation of IR, as shown. (Source: Adapted, with permission, from S.H. Schneider, "Climate Modeling," *Scientific American*, vol. 256, p. 78. Copyright © 1987 by Scientific American, Inc. All rights reserved.)

when coupled to an oceanic model. These concerns for conservation are related to the concern, discussed in the section "Atmosphere-Ocean Subsystem" (below), about numerical drift apparent in coupled atmospheric-oceanic GCMs.

Future changes in the earth system will probably result from increasing emissions (and atmospheric concentrations) of greenhouse gases such as carbon dioxide, chlorofluorocarbons, methane, and nitrous oxide (see Logan et al., 1981; OIES, 1985; Schimel et al., 1989). These substances have biological, industrial, and other anthropogenic sources. In addition to the direct radiative effect, some of these gases undergo atmospheric chemical and photochemical transformations that alter the natural balance of other atmospheric gases. For example, the changes in ozone levels in both the stratosphere and the troposphere are a source of serious global concern.

Tropospheric ozone, which is toxic to plants and animals and also acts to enhance the greenhouse effect of the atmosphere, has strongly increased in recent decades. This trend is believed to result from increasing emissions of the oxides of nitrogen and volatile organic compounds as well as methane and perhaps carbon monoxide. As a result, the "oxidation capacity" of the atmosphere has been modified, as has the atmospheric lifetime of greenhouse gases such as methane and certain halocarbons. Clearly, the chemistry of the atmosphere plays a major role in the earth system, and models attempting to describe this system must include a chemical-photochemical component.

Models that incorporate atmospheric chemical processes provide the basis for much of our current understanding in such critical problem areas as acid rain and photochemical smog production in the troposphere and depletion of the ozone layer in the stratosphere. The very formidable nature of the problems requires that the models not only include chemical processes, but also dynamical and radiative processes, which through their mutual interactions determine the circulation, thermal structure, and distribution of constituents in the atmosphere (i.e., requires a coupling of the physics and chemistry of the atmosphere). Furthermore, the models must be applicable on a variety of spatial (regional-to-global) and temporal (days-to-decades) scales.

The traditional tool in atmospheric chemistry studies for many years has been a one-dimensional model. Current versions of one-dimensional models include a large number of chemical reactions (often exceeding 100) coupled with a radiative transfer model. Vertical transport is modeled through incorporation of an eddy diffusion term, not unlike the earlier ocean carbon one-dimensional models (e.g., Oeschger et al., 1975). These models are often incorrectly interpreted as "globally averaged" models in some sense. This interpretation arises because the eddy diffusion coefficient is generally inferred from a well-mixed, long-lived tracer distribution. However, these models are generally most applicable to mid-latitude regions. The virtue of these models resides in their ability to provide long-term (decades) simulations with modest computational resources. The inability to model horizontal transport severely limits the utility of these models for future research. Progress in the development of two- and three-dimensional models and availability of more powerful computers have also negated some of the advantages of the one-dimensional model. However, these models will undoubtedly continue to be used for many years as a quick and convenient means of examining the effect of new reactions, changes in spectroscopic data, or reaction rates and alternate scenario studies.

More recently, atmospheric chemistry studies have relied upon two-dimensional models. These models solve the zonally averaged momentum, thermodynamic, and mass continuity equations, including detailed treatment of chemical and radiative processes. Because of the increased computational requirements, many such models group related constituents into "families" to avoid explicit integration of a mass continuity equation for each individual chemical species (not unlike the grouping that occurs in ecosystem models). These models must include the effects of horizontal transport by zonally asymmetric motions (waves or eddies) by eddy diffusion terms analogous to the approach adopted for vertical transport in the one-dimensional models. As a consequence of these approximations of the actual physics, these models do not correctly represent the interactive behavior of the chemical, radiative, and dynamical processes. Despite such shortcomings, these models have provided significant additional insight into atmospheric chemical processes through incorporation of horizontal motions. They will necessarily provide the basis for ozone assessment studies for at least several years until significant progress is made in developing three-dimensional models and acquiring more powerful and affordable computing resources (e.g., Bass, 1980; Demerjian, 1976).

Three-dimensional global transport models for chemically active species have been under development for the global atmosphere since the early 1970s. The first such models considered the fate of long-lived gases such as nitrous oxide or the CFCs and emphasized the improvement of the transport formulation associated with GCMs. More recently, attempts have been made to develop more chemically intensive three-dimensional global models. Although efforts to include chemistry in a three-dimensional model date back at least two decades, progress has been relatively slow owing to the enormous computational requirements for treating the fluid dynamic equations alone. The additional burden imposed by incorporating detailed chemistry into a comprehensive GCM makes long-term simulations impractical. Current three-dimensional atmospheric chemistry models seek a compromise solution by some combination of the following: adopting a relatively coarse resolution (in both the vertical and the horizontal dimensions); incorporating constituents by families (similar to the practice used in most two-dimensional models); omitting or simplifying parameterizations for tropospheric physical processes; and conducting "off-line" transport simulations in which previously calculated wind and temperature fields are used as known inputs to a series of mass continuity equations including chemical source-sink terms. This last approach renders the problem tractable and has provided much progress toward understanding the transport of chemically reacting species in the atmosphere. But this progress has come at the cost of neglecting the interactive feedback between the evolving species distributions and the atmospheric circulation, and so this approach is insufficient for predicting aspects of global change (see NRC, 1984).

A Modeling Strategy: Prognosis for Progress

Considerable progress is still needed in the development of uncoupled atmospheric circulation and chemistry models, and it should be recognized that efforts to develop a fully coupled atmospheric GCM with linked chemistry and physics are still in their infancy. The major hope for attaining this goal within the next decade lies in the concomitant developments in massively parallel processing computers and our basic understanding of the complexities at the interface between the chemistry of the atmosphere and the physics of the atmosphere. The heterogeneous reactions that produce the ozone hole are but one reflection of these complexities.

One of the major challenges is to incorporate heterogeneous reactions into models. Compelling evidence now exists that during antarctic springtime in recent years reactions on the surface of polar stratospheric cloud particles have been instrumental in the destruction of polar ozone. Much more research in laboratory kinetic studies is required to determine the reaction rates to be used in model simulations. These reaction rates are functionally dependent on available particle surface area. Microphysical models that can describe the growth rate and size distribution of the cloud particles are thus an integral requirement for including heterogeneous reactions in atmospheric chemistry models. If particle growth is sufficient, sedimentation of particles can occur, resulting in denitrification and dehydration, which contributes further to the perturbed chemistry occurring in the polar vortex. Additional observations would be extremely useful to provide a basis for developing a parameterization for this process.

The possibility of heterogeneous reactions occurring on sulfate aerosol particles at mid-latitudes has recently been suggested. Of particular concern are situations associated with enhancements of the stratospheric aerosol burden by large volcanic injection events (e.g., El Chichon). Although a substantial aerosol data base (of both satellite and ground-based data) exists, global atmospheric chemistry models typically do not include aerosol effects.

In addition to the issues directly related to chemistry, there are a number of shortcomings in current models that are related to dynamical processes and thereby affect the ability of the models to predict the distribution of chemically reacting species. For example, global models typically do not simulate those equatorial wave modes (Kelvin and Rossby gravity waves) that are thought to force the semiannual and quasi-biennial oscillations in the stratosphere. This inadequacy of the models is a result of either insufficient resolution or failure to include properly tropospheric convective processes believed to be the source of the Kelvin and Rossby gravity waves. Some atmospheric chemistry models (notably two-dimensional models) have attempted to include these effects by ad hoc methods.

Gravity waves that propagate into the mesosphere represent another example of an important atmospheric wave that global models fail to resolve. These waves deposit momentum as well as contribute to small-scale mixing of constituents in the mesosphere. Most global models that extend into the mesosphere incorporate only a simplified parameterization to represent gravity wave drag.

Finally, current atmospheric models suffer from serious deficiencies that detract from their use to study atmospheric chemistry. A critical deficiency is the inadequate treatment of cloud processes and the hydrological cycle in general. This deficiency, which is discussed more fully below, is a result of both a shortage of computational resources and an incomplete understanding of the hydrological cycle, which results in poor simulations of the observed water distribution.

In sum, there is an emerging consensus that three-dimensional models (and two-dimensional models as well) may require significantly higher resolution than is now commonly used. Furthermore, there is a need for long-term simulations (tens of years) to examine the interannual variability of both climate and chemical properties exhibited by the models and the degree to which that variability is consistent with observed statistics. These considerations pose resource demands in order to achieve progress. Additionally, there are many gaps in our fundamental understanding of chemical, physical, and dynamical processes that inhibit progress in modeling climate and atmospheric chemistry and their coupling.

Research Priorities

Much progress has been made in recent years in the development of GCMs and the use of chemical models to understand key processes in the troposphere, stratosphere, and mesosphere. But more sophisticated physical parameterizations and chemical and transport codes are required as important components of earth system models and should be developed. In this section the submodels or components needed to make further progress in simulating the chemistry of the atmosphere are briefly discussed. The focus is primarily on the chemistry, although some key issues regarding GCM development are presented.

Because of the uneven distribution of emission sources at the surface of the earth and the role of meteorological processes at various scales, models of chemically active trace gases in the troposphere should be three-dimensional and resolve transport processes at the highest possible resolution. These models should be designed to simulate the chemistry and transport of atmospheric tracers on a global and regional scale, with accurate parameterizations of important subscale processes. It is therefore necessary to have an ambitious long-term perspective to develop comprehensive models of the tropospheric system, including chemical, physical, and eventually biological components. The development of such models and their integration in even more complex earth system models will require stable long-term support for interdisciplinary research teams. A large effort will have to be devoted to studies of various individual processes, with an emphasis on those that affect atmospheric chemistry on the global and regional scales.

Models of the critical components of the tropospheric system will provide the basis for the three-dimensional chemistry-transport models. Important examples of such component models are the following:

- Models for biological and surface sources. Three types of models of the bottom boundary condition are needed: (1) Global empirical models of surface emissions are necessary to extrapolate and interpolate individual measurements provided in different environments under different conditions. These models will be based on empirical relations accounting for the variation in emissions with climate parameters such as temperature, solar radiation, and soil moisture. (2) Models of detailed biological mechanisms associated with trace gas emissions from soil, vegetation, and oceans need to be developed. The processes to be considered will range from a leaf of a tree to an entire ecosystem. (3) Models of surface exchanges and transport in the boundary and surface layers need to be used to describe the transfer between the ocean or land surfaces and the atmosphere. These models will also simulate the physical mechanisms that are responsible for the deposition of gases on different types of surfaces (see Bolin and Cook, 1983; IPCC, 1990; Lenschow and Hicks, 1989; NRC, 1984; Schimel et al., 1989).

- Models of long-range transport. Transport models driven at high temporal resolution by "weather systems" generated by GCMs are required to simulate how advection, turbulence, and convection affect the chemical composition of the atmosphere. Several numerical approaches can be used, including Eulerian and Lagrangian formulations. These models will be used (1) with minimum chemistry, to simulate the global distribution and variability of long-lived species including water vapor and emphasizing exchanges with the stratosphere and between the hemispheres and (2) with more detailed chemistry, to explore the role of meteorological processes in determining the spatial distribution and temporal variability of short-lived species. In this second case, the importance of the effects of continental pollution on the remote troposphere and on the oxidizing capacity of the atmosphere will be studied (Bass, 1980; Logan et al., 1981; NRC, 1984).

- Models for photochemical transformations. Chemical models with a detailed set of reactions, in which the transport is ignored, need to be developed. These models will describe the complex relationships between hydrogen, nitrogen, and oxygen species as well as hydrocarbons and other organic species. It is important to note that atmospheric radiation models must be developed to give the spectral distribution of solar radiation consistent with physical processes of atmospheric climate models; dependences on solar geometry, distributions of water vapor, ozone, and aerosols, surface albedo, and cloud optical properties must be treated, to provide the photon inputs needed to drive photochemical reactions. These models will be used to establish simplified chemical schemes that will be implemented in chemistry-transport models (e.g., Demerjian, 1976).

- Models of removal processes. In order to simulate the effects of clouds on the chemical constituents and to investigate the mechanisms involved in the wet removal of atmospheric constituents, models should be developed to account for the following processes: (1) cloud convection, which provides an efficient mechanism for vertical transport in the troposphere; (2) meteorological transport through the boundary and surface mixed layer; (3) aqueous transformations of species in clouds; and (4) precipitation of trace gases and wet deposition (NRC, 1984).

- Models of troposphere-stratosphere exchange. The concentrations of water vapor in the stratosphere are needed for modeling both photochemistry and atmospheric thermal infrared radiation fluxes. Thus it is crucial to establish what determines this concentration for present and changing climate conditions. It is necessary to include both transport and the photochemical source of methane oxidation. Transport can occur both from large-scale motions and from convective transport, especially at the tops of cumulonimbus towers. The intriguing similarity between observed stratospheric water vapor concentrations and saturation vapor pressures at the temperature of the climatological tropical tropopause has long led to the suggestion of a "coldtrap" mechanism. That is, if much of the mass exchange from troposphere to stratosphere moves through the tropical tropopause, formations of cirrus clouds will squeeze out water beyond saturation. Suggestions that oxidation of (increasing) methane adds to stratospheric water vapor concentrations can only be quantified in the context of models of transport exchange between troposphere and stratosphere.

- Models of the middle atmosphere. Models of the middle atmosphere are used in relation with the ozone budget. A large number of studies are based on two-dimensional models that include a relatively detailed chemical scheme as well as a radiative code that takes into account the important coupling between radiative (temperature) and chemical processes. Three-dimensional models need to be developed and applied to the middle atmosphere. These models will be able to explicitly reproduce the propagation of planetary waves and simulate the effects of these waves on the meridional transport of the trace gases such as ozone. These models are also needed to study the dynamical and chemical mechanisms involved in the formation and dissipation of the ozone hole over Antarctica. Models of the stratosphere should include the effects of aerosols and polar stratospheric clouds on chemical budgets.

- Models of cloud processes. Models of cloud processes on the small scale to the mesoscale should be developed to parameterize the transport of trace species including water vapor and the feedback between clouds, their radiative properties, and large-scale circulation.

Summary

Further progress in the modeling of atmospheric chemistry requires a better knowledge of several factors: (1) Surface sources of trace gases, in particular emissions by the biosphere, will be better understood through systematic observations of emission fluxes by different ecosystems under variable meteorological conditions and by the development of ecosystem and surface models that will provide parameterizations of these emissions. (2) Chemical processes and reaction channels, both in the gas and in the aqueous phase, will be most usefully investigated in accelerated laboratory studies. (3) Transport processes by advection and convection will be illuminated through the development of high-resolution transport models coupled to atmospheric GCMs with detailed representation of physical processes including cloud formation and boundary layer transport.

Further progress in the modeling of atmospheric physical processes depends in part on carrying out simulations at higher resolution. This is achieved through the acquisition of more powerful computational resources through advancing technology and better funding. Some progress is also possible through improvements in numerical and programming procedures and the use of finer meshes over limited areas of the globe.

Improvements in the treatments of model processes, especially those involving clouds and the hydrological cycle, surface energy exchanges, and interactions of these with radiation, are also crucial for further progress in modeling the atmosphere. These improvements must be paralleled by the acquisition of global data sets for validation of these treatments. Validation of models against global and regional requirements for conservation of energy is especially important in this regard. Feedbacks to and from the land require careful attention to the treatments of evapotranspiration, soil moisture storage, and runoff. All these occur on spatial scales fine in comparison with those of the model meshes, and so the question of scaling must be addressed. Other key developments are the treatment of subgrid-scale convection in the models including the planetary boundary layer with moist processes and the treatment of cloud and precipitation physics to provide radiatively important parameters such as cloud liquid water and drop size distributions.

Future progress will depend not only on available computational resources but also on advances in understanding fundamental physical and chemical processes and the nature of their coupling. In more direct words, progress will be determined, in part, by the scientific talent that is focused on the issues of global change.

ATMOSPHERE-OCEAN SUBSYSTEM

The challenge is to develop sufficient understanding of the coupled atmosphere-ocean system, including biogeochemical components, to allow well-reasoned prognostic statements about the dynamics of this system. Of particular interest is how the atmosphere-ocean system affects and is affected by changing global phenomena, such as climate, on interannual to decadal to centennial time scales. For example, in the context of a possibly changing climate under a given scenario of anthropogenic production of carbon dioxide what can be said about:

- the oceanic uptake and transport of carbon dioxide;
- the oceanic uptake and transport of heat;
- the distribution of new and total oceanic primary production; and, finally,
- what will be the impacts on, and feedbacks from, the ocean as the result of a changing hydrological cycle?

Answering such questions will require an expansion of the comprehensiveness of the processes that current models incorporate explicitly or implicitly through parameterization. In addition, in some cases, parameterizations need improvement; in others, previously parameterized processes may need to be resolved or previously resolved processes may lend themselves to parameterization. There are also issues regarding the coupling of oceanic, atmospheric, and biogeochemical models, including the problems of possible mismatches of time and space scales.

The process for substantially improving current models as well as creating new models should recognize the need for exploration, evolution, and close interaction with major field programs (see chapter 7). The development needs to be geared toward early tests and trial attempts. In fact, some "great failures" early in the effort would be very instructive in guiding model development not only in the atmosphere-ocean system but throughout the earth system modeling effort. Fortunately, the foundation—the existing models—for this development is both broad and deep. These existing models include uncoupled and coupled models.

Uncoupled Models

Uncoupled models of oceanic circulation, atmospheric circulation, and biogeochemical-biological processes are being employed and improved. These three classes of models have varying degrees of maturity and sophistication. Perhaps the greatest degree of development has been reached by atmospheric GCMs, in which simulations with prescribed oceans (and other boundary conditions) replicate many of the observed features of weather and climate (see the section "Atmosphere-Terrestrial Subsystem" above). Similarly, although oceanic GCMs are somewhat less developed than their atmospheric counterparts, oceanic simulations in which air-sea exchanges are prescribed replicate many aspects of the large-scale characteristics of ocean temperature, salinity, and circulation. Least developed are biogeochemical-biological models in which physical processes such as fluid flows and mixing are prescribed from either theory or observations and ecosystem dynamics develop under forcing and controls of sunlight, temperature, and nutrients. These models nevertheless reproduce the evolution in time (e.g., seasonal changes but not interannual trends) of biological populations and the corresponding rate of primary and secondary production. However, they tend to deal with lumped populations (phytoplankton, zooplankton, and so on) rather than biogeochemical state variables (such as total dissolved inorganic carbon and alkalinity) necessary for issues such as the carbon cycle.

Until recently, uncoupled oceanic GCMs have come in two distinct varieties: (1) basin-wide to global, coarse-resolution models that fail to resolve mesoscale eddies and (2) eddy-resolving models that are confined to limited domains. Now these two varieties of GCMs are converging. Advances in computational power and improvements in coding have allowed global oceanic GCMs to achieve sufficiently fine spatial resolution to begin to resolve mesoscale eddies.

This convergence between "coarse scale" and "eddy resolving" oceanic models is similar to that occurring between "weather prediction" and "climate" models, as noted in the section "Physical-Chemical Interactions in the Atmosphere" (above). High-resolution models used in short-term weather forecasts now incorporate processes and components that are traditionally the purview of climatology (e.g., soil moisture and other parts of the global hydrological cycle) in an attempt to extend the useful range of forecasts;

coarse-resolution climate models are beginning to utilize enhanced spatial resolution and account for additional processes (e.g., topographic interactions with evapotranspiration).

Uncoupled marine biological and biogeochemical models with prescribed atmospheres and oceans are in the beginning stages of development and again in the two typical forms: (1) localized, high-resolution and (2) lower-resolution extended-area versions. The former tend to incorporate more complex ecosystem representations, whereas the latter focus on simplified biogeochemistry in basin-wide or global domains. There are, however, some recent rather encouraging attempts (e.g., the joint work at Princeton University and the Geophysical Fluid Dynamics Laboratory (GFDL); see Sarmiento et al., 1988; Toggweiler et al., 1989; and the work in Hamburg; see Maier-Reimer and Hasselman, 1987) at adding biological complexity to global treatments.

There are several examples of high-resolution localized biological models in which a specified flow field is used to advect and diffuse changing biological populations. Models of this type generally are formulated and implemented to study processes that occur over limited space and time scales in specific coastal regions and rely upon synoptic distributions of such properties as mixed-layer depth and phytoplankton distribution. At present, they are sufficient for investigating processes; they are not at the level of sophistication or resolution needed for attempting prognostic investigation.

Interestingly, the addition of simple (though increasingly complex) biological-biogeochemical models into oceanic GCMs could allow for the study of possible biological changes under a forcing of changing climate and oceanic circulation fields. Of particular interest would be the species shifts and other ecological transitions that likely would occur in a changing climate; these issues are not as yet considered in this complex modeling environment (within a GCM), and hence changes in the biogeochemical system that might result from ecological changes are not treated. Also in this context, although there is no true coupling (the physics drives the biology— there is no feedback from the biology such as effects on mixed-layer depth), the structure certainly has the potential to allow a biological feedback on the circulation.

These biological effects are important, since simple box models indicate that changes in oceanic biogeochemistry can profoundly influence the atmosphere-ocean partitioning of carbon. Addressing future changes in oceanic biogeochemistry is essential to bracket adequately the oceanic uptake of carbon dioxide. This capability requires understanding the dependence of biological-biogeochemical processes, especially new production and decomposition and carbonate formation and dissolution, on current and future circulation patterns. Currently, we cannot even specify the sign of the feedback.

Coupled Models

Coupled three-dimensional circulation models of both the ocean and the atmosphere have been under development for several years. Recently, some exploratory attempts have been made to study the response of the model coupled climate system to realistic time-evolving scenarios of the greenhouse forcing. However, coupling oceanic and atmospheric models introduces new problems of unrealistic long-term drift and instabilities that are not exhibited by the uncoupled models. Further, adding biogeochemical processes to coupled oceanic-atmospheric GCMs is a prerequisite for complete description of the effect of carbon dioxide removed from the atmosphere by the ocean. Adding biology and chemistry to GCMs is a nontrivial exercise. The complexity of the fine-scale response of uncoupled ecological models, if proven necessary, could potentially overwhelm available and foreseen computing resources.

Coupled abiotic oceanic-atmospheric global GCMs are beginning to be used to project the delay in the climate's response to greenhouse forcing due to the thermal inertia of the oceans (e.g., Bryan et al., 1982; Cess and Goldenberg, 1981; Manabe and Wetherald, 1975). However, the conclusions are still uncertain to within a factor of at least 2 (IPCC, 1990). Furthermore, the pattern of oceanic response (e.g., which hemisphere warms faster) is also model dependent. The use of coupled oceanic-atmospheric general circulation models (O/A-GCMs) in short-term forecast and data assimilation models is expanding; this will probably put prognostic climate simulations on a firmer foundation.

Coupled models of the oceans and biogeochemical systems are beginning to be explored. Particularly important in this exploration is the use of satellite-derived ocean color data (see chapter 7) and the consideration of the biological controls on mixed-layer depth and thermal distribution in the surface layer of the ocean. For example, vertical changes in the water column heating rate due to the vertical gradient of phytoplankton can potentially alter vertical mixing intensity and the depth of the mixed layer. Inclusion of such processes could be important in earth system models focused on global change on decadal time scales since such processes provide a feedback from the biological system to the physical system. Also, changes in the intensity of solar irradiance constitute an important aspect of this physical-biological interaction, which in turn couples the oceanic physical-biological model to atmospheric processes. Most of this work, thus far, is in the form of either preliminary calculations using satellite data in order to determine the importance of the phenomena at global scales or generic mixed-layer models that are parameterized using oceanic data for specific regions (e.g., Bermuda), thereby producing simulations of biogeochemical cycles on both global and regional scales. Incorporating into the models descriptions of the oceanic surface state (wave models) will improve the calculations of air-sea carbon dioxide flux calculations.

Another theme in modeling a coupled ocean-biology system that merits further exploration is the use of physical and chemical tracers coupled with a set of independent constraints (e.g., geostrophy and poleward flux of heat) to infer, through a matrix inversion or linear (or quadratic) programming technique, the rates of oceanic motion and biological activity that are consistent with these tracer fields (e.g., Bolin et al., 1983; Moore et al., 1989a; Wunsch, 1978; see also Broecker and Peng, 1982; Riley, 1951). The resulting model is obviously not prognostic, and its coupling between biology and physics is but a simultaneous or perhaps statistical parameterization, but the approach has great value in that it yields a diagnostic tool for use in an active interplay with efforts to incorporate biological-biogeochemical models into oceanic GCMs.

Finally, coupled models of the oceans, atmosphere, and the biology-biogeochemistry are in the beginning stages of development for at least local and regional domains. Such limited-area models, of course, are incomplete in the sense that boundary conditions at the sides of the domain must be externally specified. (This consideration also restricts the useful time span covered by such models to just one or two weeks.) Theoretically, such models could be extended to cover much larger—even global—domains. It is an ambitious but not unrealistic goal to have a fully coupled, prognostic (highly simplified), three-dimensional oceanic-atmospheric GCM including biological-biogeochemical dynamics running by the year 2000.

A Modeling Strategy: Prognosis for Progress

There are traditionally two approaches to a modeling strategy; in jargon, they are called "bottom-up" and "top-down." The former seeks to improve models by improving specific processes through explicit process-specific experimentation or data recovery. The latter studies the macroresponse of models to major perturbations or phenomena and thereby seeks to evaluate and hence improve models.

Bottom-Up: A Focus on Key Processes

The realism or quality of today's generation of oceanic GCMs is limited by a variety of numerical and dynamical factors. Among these are (1) an inadequate theoretical or observational understanding of certain key processes fundamental to the earth's coupled climate and biogeochemical subsystems and a corresponding and continuing uncertainty as to how best to incorporate or to parameterize them in oceanic GCMs and (2) a need for improved numerical methodologies (both prognostic and assimilative), for the use of enhanced resolution within the models, and for the computational and observational resources necessary to run and to validate the models. Here, the committee attempts to identify those physical processes that, although important to the ocean's role in climate and biogeochemical systems, are nonetheless poorly understood from either a theoretical or an observational perspective.

In making this identification, the committee has tried to categorize crudely the processes according to their characteristic (horizontal) spatial scale; the three categories used here are (1) small scale (10 km or less in the horizontal or 10 m or less in the vertical), (2) mesoscale (from the grid scale of the oceanic GCM, assumed to be 10 km to a few hundred kilometers), and (3) basin and global scale. The resulting "small-scale" processes are by definition therefore likely to be unresolved by any foreseeable GCM system, but their effects on larger-scale phenomena will nevertheless need to be incorporated via adequate parameterized terms in the oceanic equations of motion. The processes labeled "mesoscale" are those whose spatial extent is subbasin scale, resolvable by the numerical model grid, and known (or presumed) to be crucial to the form and strength of the global circulation. Likewise, processes operating on the largest spatial scales have been termed "basin and global scale," and discussion of these shifts to a "top-down" focus.

Key Processes: Small Scales The small-, or subgrid-, scale processes, which need to be much more thoroughly understood and, ultimately, better parameterized, can be broken into two categories: (1) those that have both physical and biological consequences and (2) those that are strictly biological in their implications. In the former category are, for instance, horizontal and vertical diffusion, and small-scale convective processes; and in the latter are a wide variety of biological-biogeochemical interactions including grazing, photosynthesis, and microbial processes. Examples of questions include the following: What processes determine the various temporal and spatial scales of tropical convection? How do the finest scales interact with the mixed layer? Does turbulence affect location and identification of prey by grazers? Is microscale patchiness in nutrients important in regulating nutrient incorporation by phytoplankton?

Small-scale processes are important for one simple reason: their effects are biased. Thus the mean of the small scales over sufficient length and time scales is significant. A recent workshop on subgrid processes showed the dependence of vertical and horizontal fluxes on the nature of the small-scale parameterization. This bias may be taken advantage of when our interest lies in the longer space-time scale and we have sufficient data to parameterize the effect.

The bias of turbulence is that in general small-scale processes are cascading to even smaller scales, which then are removed by friction. The small scale must be related to the large because the source of the small scale is the large scale and because scales will generally decrease. The reality is that coherent turbulent structures exist and at times small scales are combined to form larger scales. This implies, in effect, a negative eddy viscosity (diffusivity).

The biological effect of small scales is seen in the nonlinear response of phytoplankton and zooplankton to light and nutrient concentrations. The result of such nonlinearity is that the average response of a system (biological communities) to a "patchy forcing" (by light or nutrient concentrations) is different from the response of the system (community) to the average forcing.

In sum, bulk formulations seem adequate to a first approximation, but better parameterization of small-scale processes may be needed. By their very nature, all these processes are difficult to observe and model, being characterized by very small spatial (and sometimes temporal) scales. While the potential for small-scale effects has been identified, further progress is required in order to quantify the effects. The physical community is still working toward adequate parameterization of the eddy diffusivity, and the biological community is beginning to consider the effect of patchiness. Learning how to deal better with such processes, in the sense of providing adequate parameterizations or functional descriptions of them, will be a continuing concern; however, progress in these areas can be made by conducting carefully designed oceanic and laboratory experiments to observe and quantify small-scale physical and biological processes. One example of the type of experiment necessary is the so-called tracer release experiments, whose goal is to measure isopycnal and diapycnal mixing rates in the ocean.

Limiting Processes: Mesoscales Over the past decade, much progress has been made in identifying and understanding several mesoscale processes that profoundly influence the oceanic circulation on either a regional or a global basis. The list of processes for which a preliminary, though far from complete, understanding is available includes mesoscale eddy dynamics, meandering jets, fronts, and upwelling. However, the relationships between mesoscale eddy dynamics and biological productivity are poorly understood. Despite this initial progress, however, other important phenomena on this scale remain incompletely explored. These include the formation and spreading of bottom and intermediate-depth water masses, and the effects of synoptic-scale atmospheric fluxes and forcing.

Fortunately, on this scale, further understanding of processes—including their respective roles in the climate and biogeochemical subsystems and their geographical and parametric dependencies—can be achieved as well by process-oriented numerical modeling as by direct observation. Both will be valuable and should be encouraged. For example, although much remains to be learned about how best to model regional air-sea coupling and water mass formation processes, models do exist that are capable of preliminary, but sophisticated, exploration of such effects. Likewise, future observational efforts (e.g., the Office of Naval Research's Oceanic Subduction Experiment, and TOGA/COARE; see chapter 7) will contribute the observational basis for validating current mesoscale and basin-scale oceanic models.

Mesoscale processes are particularly troublesome since they are too large to be safely ignored and yet the dynamics of biology and physics are not well understood at this scale. Further study employing both models and observations is needed to define which processes can be reasonably treated through parameterization and which must be modeled explicitly.

Limiting Processes: Basin and Global Scales The behavior of the global climate and biogeochemical systems is clearly tied to the nature, structure, and properties of the global oceanic circulation and primary production. On its largest scales, the oceanic general circulation comprises several interacting elements, none of which is understood with the level of observational or theoretical understanding necessary, in the long term, to develop adequate global models. Of particular importance on the basin and global scales are intense boundary current systems (on both the eastern and the western boundaries), basin-scale oceanic gyres, coastal/deep ocean interaction and exchange, ice dynamics, and deep abyssal circulation and its relationship to ocean topography. Perhaps the most important biogeochemical processes are those that regulate the "efficiency" of the biological pump in high-latitude waters: Why is surface nitrogen trioxide always so high in the Southern Ocean? Here again, a mutual interaction among process-oriented numerical experiments, global-scale observational programs (e.g., WOCE, JGOFS, and GLOBEC), and assimilative modeling is needed to fill out our understanding of global exchange processes to the level necessary for the testing and validation of coupled climate models.

Top-Down: A Focus on Phenomena

In the top-down approach the strategy is one of extracting information from large-scale natural phenomena. There are several phenomena that are crucial to the coupled atmosphere-ocean-biogeochemical system. These phenomena are large-scale events or processes that should be accommodated in future global change models and will act as specific tests of the model dynamics (see the section "Critical Model Tests" (below), where specific global change tests are discussed). Such model tests should also play a critical role in developing hypotheses and experimental designs for observational programs focusing on these specific phenomena. As an outcome of such programs, large-scale data assimilation may be feasible and may be particularly important as part of model evaluation. Furthermore, studies of these large-scale processes will serve not only as diagnostic tests but also as prognostic tools. In order to illustrate the range of these phenomena, a few of them are described here:

- El Niño/Southern Oscillation (ENSO). The modeling activities surrounding this phenomenon could serve as a standard for other large-scale, phenomenon-oriented modeling efforts. Significant progress has been made in ENSO models, especially in terms of the coupling of atmospheric and oceanic dynamics (the TOGA program has been central in this development), but the coupling between physical and biogeochemical processes such as new primary production and carbon dioxide exchange has not been modeled, although work is in progress. Because of the very large changes in the atmosphere and ocean (both physically and biologically) that occur in conjunction with this phenomenon, it offers an ideal arena for validating physical and biological models.
- Poleward heat flux. Poleward heat flux in the ocean and atmosphere is crucial in the earth's heat budget. Large-scale, "steady" transport is fairly well understood. However, the uncertainties are large, and the magnitude (as well as the sign) of the flux is extremely sensitive to the parameterization of the eddy fluxes. Other processes such as the "eddy-Ekman" flux (the interaction between the zonal wind and ocean surface temperature variability) are not well understood. The challenge for modeling will be to resolve these processes as well as the time delay between oceanic heat uptake and release and changes in atmospheric circulation.
- Spring bloom. One of the difficulties with incorporating biological processes in oceanic GCMs (aside from the scale differences) is that biological patterns often do not evolve smoothly; rather, they behave like "switching processes," whereby the biological system and structures change rapidly from one state to another. The spring bloom in the North Atlantic is an example of this behavior as well as being important in overall primary production of the ocean and carbon dioxide exchange. A series of modeling efforts should be developed focusing on this region.
- Cross-shelf exchange. Processes associated with continental margin systems are continuous with those of the open ocean; furthermore, the exchanges between the margins and the open ocean (i.e., the cross-shelf exchanges) of material, heat, and momentum are important. The implementation of models (e.g., physical-biological) to estimate exchanges between the margins and the open ocean has associated with it a significant problem—the space and time scales that characterize processes, as well as the processes themselves, can be quite different in coastal and open ocean environments. For example, low-frequency, large-scale events such as the ENSO and mesoscale eddies and large-scale permanent features such as boundary currents and shelf-slope fronts can result in significant high-frequency alterations of

coastal systems with possibly major effects on the biogeochemical systems. Thus a severe test of a coupled oceanic-atmospheric-biogeochemical model will be how it resolves scale mismatches between continental margin and open ocean systems. It is likely that the initial coupling between the high-resolution coastal models and the lower-resolution open ocean component will be through bulk parameterization of the processes. The concern, then, is the need to develop realistic approaches for such parameterizations.

• Biogeochemistry of aeolian deposits. As the central gyres and the Southern Ocean have been hypothesized to be iron-limited, aeolian inputs of Asian dust and other land-derived aerosols may play a crucial role in ocean primary production. A coupled model including such a phenomenon would also need to include a component describing changes in vegetation cover and land-atmosphere interactions to provide realistic input of dust to the atmosphere and eventually to the ocean. Such a model is currently beyond our grasp, but tests of the effects of prescribed dust inputs on biogeochemical cycling could be performed.

Research Priorities

Fundamental oceanic processes important in the determination of the global biogeochemical cycles and the world's climate are poorly understood. Proper inclusion of these processes in models will require field programs tailored to understanding individual processes. The objectives of these studies will be to understand the specific mechanisms involved and to develop useful parameterizations.

The set of field programs required to acquire the data needed to advance our knowledge of fundamental oceanic processes is already well defined (e.g., see chapter 7). These programs also offer valuable opportunities for simultaneous observational efforts, and these should be encouraged. Whenever possible, observations of the atmosphere-ocean-biosphere should be made in conjunction with specific studies. In particular, oceanographic cruises frequently neglect the atmospheric marine boundary layer, biology is neglected by physical oceanographers, and biologists may minimize the physical measurements. However, the earth system is an interactive system that requires measurements of all components. Fortunately, there has been already an encouraging spirit of cooperation between TOGA, WOCE, and JGOFS. Such cooperation should be further encouraged.

From the perspective of testing atmospheric-oceanic-biological models by the use of large-scale events or phenomena, we must know the details of the phenomena even if the eventual "outcome" is to parameterize them. Characterizing the scale and magnitude of the events will constitute critical tests of model validity. The important scales are typically large. Therefore satellite measurements are important, although they are also limited. In situ measurements will be required to define the subsurface variability (satellites do not detect many key processes, for example, new production and vertical fluxes) and to provide a baseline for satellite measurements. Currently, an important concern is whether or not there will be an ocean color satellite in orbit during the major field campaigns (i.e., WOCE and JGOFS). This represents a potential critical gap in being able to link biological and physical phenomena.

Data assimilation is an emerging reality in physical oceanographic modeling and observational studies and may be used advantageously by biological and biogeochemical oceanographers. This methodology aids in the interpolation of physical observations by adding dynamical constraints. While the quantity of physical data required to describe oceanic phenomena may be reduced by the use of data assimilative models, it is more likely that field estimates will be improved as a consequence of data assimilation.

The first attempts are now being made to develop the techniques necessary to assimilate ocean color measurements into regional physical-biological models. This is a promising direction for the development of models that ultimately will have predictive capability for biological distributions in the ocean.

One aspect that makes data assimilation into physical-biological models challenging is that updating one ecosystem component (e.g., phytoplankton from ocean color) requires that all other ecosystem components be adjusted so that they are in equilibrium with the updated field. More specifically, assimilative models will require estimates of the error fields of both the assimilated data sets and the processes that are being parameterized. This will allow quantitative estimates of the confidence in the forecast (or hindcast) fields being produced. For example, in assimilating ocean color data into a multicomponent ecosystem model, one needs to have an estimate of the errors in the satellite data in time and space (i.e., particularly those associated with gap filling) as well as an estimate on the effect of zooplankton grazing within the model. Such error estimation will require synergy between the modeling effort and the field programs.

In spite of these difficulties, the initial attempts at assimilating ocean color data into physical-biological models have shown that the accuracy of the model is improved, but that the improvement of the model diminishes after a short time. The implication that data are needed at frequent intervals for assimilation into physical-biological models is a potential area of research that could be an important aspect of developing models to address problems of carbon dioxide uptake by the ocean. Specifically, given the inherent nonlinearity of biological processes as well as their occasional "switching circuit" behavior, present data assimilation techniques are inadequate. Research into techniques involving nonlinear and nondifferentiable forms is needed.

From a more classical approach, biological models in which the flow field is set as a boundary condition have been in use for about 15 years for specific regional studies. Consequently, the dynamics and limitations inherent in these models are beginning to be understood. If physical-biological models are to be developed to address the larger question of carbon dioxide uptake by the ocean, then the question arises as to how to extend the knowledge gained from regional, physically forced biological modeling studies—as well as from geographically restricted fully coupled models—to models developed for basin or global domains. While, in principle, it seems straightforward to simply increase the model domain, in practice, this is not so. At least three issues must be addressed: (1) how to link the biological dynamics to biogeochemical changes important for global carbon studies, (2) how much of the complexity that characterizes coastal biological systems needs to be transferred to larger-scale domains, and (3) how to match the space and time scale requirements of coastal processes with those of larger-scale systems. These three issues represent fundamental problems that must be addressed if coupled oceanic-atmospheric-biogeochemical models are to be developed to investigate carbon uptake by the world oceans.

Biogeochemical models link biology and chemistry at the level of nutrients and carbon dioxide and are generally based on the forcing of nitrate or phosphate fields. The full effect of the biology on the chemistry is generally not included, nor is the full effect of the chemistry on the biology. Yet it is likely in a changing climate system that these processes may be important. We need to better understand the sensitivity of the climate system to changing biogeochemical systems, and, if found to be relevant, biogeochemical systems must be included in our climate models. Including these relationships is expected to be computationally expensive, and yet ignoring the interaction of the physical-chemical-biological systems may lead to poor predictions of climatic change.

Local to regional three-dimensional, coupled oceanic-atmospheric-biological models that use circulation fields obtained from sophisticated regional primitive equation circulation models to produce "predictions" of biological distributions are currently under development. The development of this type of model should be given particular encouragement since this development is essential for the realization of fully coupled global carbon cycle models. In particular, these regional models can be used to test and validate parameterizations and generalizations that are used in larger-scale models. The development of realistic fine-scale regional models is also desirable in that the output from these models can be used to specify the boundary conditions for the basin- and global-scale models. Perhaps the appropriate direction for modeling is along two parallel paths: one to develop basin- and global-scale models with increasing levels of coupling and the second to develop a series of regional fine-scale models that could provide boundary conditions and parameterizations tests for the larger-scale models. Each regional model could, thereby, include the complexity and dynamics appropriate for simulating the processes in a specific region, and at the same time the necessity for maintaining reasonable and consistent interfaces with the basin- and global-scale models would give the entire modeling effort an overall framework.

Finally, much can be learned from extant models despite their limitations. Careful analysis of the sensitivity of the systems, which are approximations to the climate and biogeochemical systems, will indicate the emphasis needed in observational and modeling studies. The importance of experience gained through modeling experiments, including failure, should not be underestimated. Failure, when carefully analyzed in the refereed literature, can be valuable to the scientific community as a whole.

Summary

The strategy is to acquire data through field experiments (e.g., TOGA/COARE, WOCE, JGOFS; see chapter 7) designed to develop an understanding of processes and a description of phenomena. Models may aid in the design and execution of these experiments as well as in the analysis and interpretation of the measurements. New understanding of key processes will be used to improve models and reduce or improve parameterizations. These model enhancements are expected to lead to the ability to describe biogeochemical and physical phenomena. This enhanced modeling ability should lead to improved climate estimates, including error bounds from which rational decisions may be made. Hence the development of the fully coupled models should be encouraged along two parallel paths: one to develop basin- and global-scale models with increasing levels of coupling and the second to develop a series of regional fine-scale models that could provide boundary conditions and parameterization tests for the larger-scale models. Each regional model could include the complexity and dynamics appropriate for simulating the processes in a specific region, and the necessity for maintaining interfaces with the larger-scale models would provide an overall consistent structure for model development.

In seeking to develop models of the coupled atmosphere-ocean-marine biosphere and biogeochemical system, it is important, as mentioned earlier, to recognize the value of "great failure." Linking atmospheric-oceanic-biospheric models, even though costly in terms of human and computer resources, should begin sooner rather than later. These early, relatively primitive attempts will shed light on the difficult issues of scale, both spatial and temporal, and the associated questions concerning the degree to which various process complexities or details are required. This clarification may be of particular importance as we scale the biological-biogeochemical components from local-regional domains to basin-global domains, or as we seek to better define and formulate the upper mixed-layer physics and possible biological feedback, such as shading due to phytoplankton blooms.

At the least, such attempts will encourage the creation of needed infrastructure and will provide a basis for assessing better the required resources. Part of the infrastructure enhancement would be the establishment of modeling teams. Obviously, several parallel efforts will be needed.

Validation of these models is both difficult and critical. The first step is to ensure that they reproduce major climate phenomena (e.g., the spring bloom and El Niño). Testing (not validating) the "interfacing" models as well as the earth system models that link them can be addressed in the U.S. Global Change Research Program.

CRITICAL MODEL TESTS

The earth system modeling program should include three interface models, as well as models of the fully coupled system. This approach allows for the rapid development of science and its inclusion into the less computationally demanding (although still challenging) interface models. Also, certain avenues of validation are open to the interface models that will be difficult to use for a full earth system model. The required abilities of the models and the critical tests needed before they can be used with confidence are discussed below. As concepts are developed and tested in the interface models, they should be included into an evolving earth system model that will form the basis for long-term prediction.

The Challenge and Critical Tests

All of the models described below must be able to simulate system response to the forcing induced, for example, by a carbon-dioxide-equivalent doubling in the atmosphere. That is, all of the models must be able to simulate the transient response of oceans, ecosystems, chemistry, or physical atmosphere to a change in physical climate induced by a greenhouse gas forcing.

Other drivers and critical feedbacks (e.g., land surface albedo, clouds, and oceanic heat transport) should be included when developing physical climate scenarios for use as forcing functions. The forcing functions given to the interface models will evolve as tested concepts from the interface models are incorporated into the earth system model, presumably modifying its predictions of whole-system response to changing greenhouse forcing. Thus a continual interplay of interface and earth system models is required, allowing for cyclic validation, failure, and modification.

Finally, validation is impossible in the classical control-experiment mode. We have no other earth system to serve as a control, not to mention the difficulties imposed by the multidecadal time frame and policy-relevant aspects of this science. As a step in appraising these models, a series of critical tests is described below that exploit various data sets, including satellite data (e.g., Earth System Science Committee, 1988) and paleo-records (see chapter 3). These "tests" of these interface models provide an evaluation of the models' capabilities prior to either their use in a predictive mode or their inclusion in an earth system model.

The Interface Models

Atmosphere-Terrestrial Subsystem

The challenge for an interface model of the atmosphere-land biosphere is to predict responses to changes in such phenomena as

- water and energy exchange (and more generally the hydrological cycle per se),
- trace gas biogeochemistry,
- primary productivity and ecosystem carbon storage, and
- vegetation composition and structure due to the changing macroclimatic forcing and/or atmospheric chemical composition.

Critical tests of this model prior to its use in the predictive mode will be to

- reproduce current patterns of biogenic trace gas and carbon exchange, using past and current climate as drivers;
- reproduce key aspects of coupling between the paleoclimate and paleoecological records within regions of interest;
- simulate contemporary spatial and seasonal patterns of vegetation properties, including primary productivity worldwide, using satellite indices as validation data;
- capture patterns of ecological change along anthropogenically induced chemical gradients in the land component of this interface model; and
- simulate surface fluxes of radiation, especially solar, and including spectral surface albedos in the atmospheric component of this interface model. Adequate simulation of amounts and spatial and temporal distribution of precipitation must also be addressed.

Physical-Chemical Interactions in the Atmosphere

For this interface model of the physical atmosphere and the chemical atmosphere, the challenge is to predict the change in chemical climate through a macroclimatic transient forcing. This will, of course, require either simulation or forcing functions for the biospheric sources, which could be derived from the atmospheric-terrestrial-biospheric model.

Critical tests will include simulation of

- contemporary variations in global trace gas fields, especially of "integrator species" such as methyl chloroform and carbon monoxide;
- methane and carbon dioxide concentrations and isotope ratios;
- large-scale tropospheric ozone features such as are observed in the tropics;
- high-latitude stratospheric ozone; and
- exchange of water vapor between troposphere and stratosphere.

Atmosphere-Ocean Subsystem

For the interface model of the atmosphere-ocean subsystem, the challenge is to predict the responses of

- water and energy exchange,
- carbon dioxide exchange and carbon storage,
- pattern of the spring bloom, and
- shifts in ecosystem composition and resultant shifts in oceanic mixed-layer chemistry (e.g., alkalinity).

The critical tests of such models are whether they can capture the key aspects of such large-scale phenomena as

- El Niño/Southern Oscillation (ENSO),
- North Atlantic spring bloom,
- cross-shelf exchange,
- poleward heat flux, and
- biogeochemistry of aeolian deposits.

INFRASTRUCTURE

The global change modeling effort, particularly on these longer time scales, encompasses a class of scientific problems far broader than those characterized by the physical climate system alone. The biogeochemical system (see Figure 2.1) merits an equal emphasis. An overall strategy that favors diversity is required, in that no one institution or group of investigators has more than a fraction of the interdisciplinary talent necessary for the complete task. Research teams in a range of sizes should be supported. Larger groups are needed for an overall integration role; smaller groups (5 to 10 people) would achieve the incremental steps (i.e., the linkages between the interface models) toward integrated earth system models. Individual investigators will obviously also make important contributions along these same lines. Some of these groups may act primarily as synthesizers whose principal interest would be in linking component pieces; others would develop the components.

The larger groups would most likely be associated with various centralized facilities, which would also serve the common needs of the various teams for computational resources and linkages with large-scale models. Examples of such needs include the preparation and analysis of large-scale observational data sets (such as those that will be developed in preparation for the NASA Earth Observing System (EOS), not to mention the essential data set that EOS will provide following launch); operation of the large, computationally intensive GCMs; documentation and maintenance of baseline codes and protocols for information exchange used by the community; and diagnostics of model output. These larger facilities would provide the physical locations for the most capable supercomputers.

One of the more intriguing advances in computer technology is in the area of massively parallel architectures. It appears that many of our current models may be recast to operate in a parallel mode. Centralized facilities could devote resources to this longer-term investment that would be difficult for smaller modeling teams to provide.

Given the rapid improvements in CPU power, especially in the workstation class and with parallel architectures, the gap between hardware and software is increasing. Many of the model codes that we use are many years old, and it is difficult to find the funds or the researchers required to convert such codes to take advantage of new hardware. In addition, many of the codes are unwieldy and poorly documented. Again, this is an area to which larger, central teams could commit resources. Specifically, more effort needs to be placed on software development than simply applying existing codes in faster machines.

To aid in this process, all modeling teams, large and small, should be encouraged to take advantage of various debugging and software development tools. For example, code profilers that aid in parallelizing or vectoring codes should be used. Object-oriented methods that allow codes to be reused or reconfigured more easily should be incorporated into new models. One of the limitations of such efforts is that such tools are often cumbersome and difficult to learn. Thus it is essential that new partnerships be formed between the various hardware and software vendors and the scientific users so that appropriate tools can be developed.

Lastly, the realm of visualization is becoming increasingly important in handling the volume and increasing dimensionality of the data sets. Visualization tools also need to be made more accessible. In addition to facilitating the analysis of the model output, such tools can play an important role in testing and debugging by allowing the modeler to see every time step of the model, rather than relying on summary data sets. Visualization also requires close coupling between the model and a data base system to track the model output.

Clearly, these centers and the smaller modeling groups and individual investigators must be mutually supporting. Smaller groups must be provided with the capability to run process experiments on full GCMs and full earth system models at larger centers; moreover, they must have on-site computer support, including workstations, advanced graphics, geographical information systems, and mainframes and, most importantly, the technical staff to allow a full application of the on-site computer facilities as well as the off-site supercomputers. Correspondingly, the centers must be able to incorporate advances in subsystem and interface representations formulated by the smaller modeling groups. Such a strategy must necessarily involve multiagency support over many years.

REFERENCES

Aber, J.D., J.M. Melillo, and C.A. Federer. 1982. Predicting the effects of rotation length, harvest intensity, and fertilization on fiber yield from northern hardwood forests in New England. Forest Sci. 28(1):31-45.

Allen, T.F.H., and E.P. Wyleto. 1984. A hierarchical model for the complexity of plant communities. J. Theor. Biol. 101:529-540.

Anthes, R.A. 1983. Regional models of the atmosphere in middle latitudes (a review). Mon. Wea. Rev. 111:1306-1335.

Bass, A. 1980. Modeling long-range transport and diffusion. Pp. 193-215 in Conference Papers, Second Joint Conference on Applications of Air Pollution Meteorology, New Orleans, La.

Bolin, B., and R.B. Cook (eds.). 1983. SCOPE 21: The Major Biogeochemical Cycles and Their Interactions. John Wiley and Sons, Chichester, England.

Bolin, B., A. Bjorkstrom, K. Holmen, and B. Moore. 1983. The simultaneous use of tracers for ocean circulation studies. Tellus 35B:206-236.

Bolin, B., B.R. Doos, J. Jager, and R. Warrick (eds.). 1986. SCOPE 29: The Greenhouse Effect, Climatic Change and Ecosystems. John Wiley and Sons, Chichester, England.

Botkin, D.B., J.F. Janak, and J.R. Wallis. 1972a. Some ecological consequences of a computer model of forest growth. J. Ecol. 60:849-873.

Botkin, D.B., J.F. Janak, and J.R. Wallis. 1972b. Rationale, limitations, and assumptions of a northeastern forest growth simulator. IBM J. Res. Develop. 16:101 116.

Broecker, W.S., and T.-H. Peng. 1982. Tracers in the Sea. Lamont-Doherty Geological Observatory, Columbia University, New York, N.Y.

Bryan, K., F.G. Komro, S. Manabe, and M.J. Spelman. 1982. Transient climate response to increasing atmospheric carbon dioxide. Science 215:56-58.

Cess, R.D., and S.D. Goldenberg. 1981. The effect of ocean heat capacity on global warming due to increasing carbon dioxide. J. Geophys. Res. 86:498-602.

Demerjian, K.L. 1976. Photochemical Diffusion Models of Air Quality Simulation: Current Status. Assessing Transportation-Related Impacts. Special Report 167. Transportation Research Board, National Research Council, Washington, D.C., pp. 21-33.

Dickinson, R.E. 1984. Modeling evapotranspiration for three-dimensional global climate models. Pp. 58-72 in J.E. Hansen and T. Takahashi (eds.), Climate Processes and Climate Sensitivity. Geophysical Monograph 29. American Geophysical Union, Washington, D.C.

Eagleson, P.S. 1986. The emergence of global-scale hydrology. Water Resour. Res. 22:6S-14S.

Earth System Sciences Committee. 1988. Earth System Science: A Closer View. National Aeronautics and Space Administration, Washington, D.C.

Emanuel, W.R., G.G. Killough, W.M. Post, and H.H. Shugart. 1984. Modeling terrestrial ecosystems and the global carbon cycle with shifts in carbon storage capacity by land use change. Ecology 65:970-983.

Farquhar, G.D., and S. von Caemmer. 1982. Modeling of photosynthetic response to environmental conditions. In O.L. Lange, P.S. Nobel, C.B. Osmond, and H. Ziegler (eds.), Physiological Plant Ecology, Encycl. Plant Physiol. (NS) 12B:549-587. Springer, Berlin.

Hanks, R.J. 1985. Soil water modeling. In M.G. Anderson and T.P. Burt (eds.), Hydrological Forecasting. John Wiley and Sons, Chichester, England.

Houghton, R.A., J.E. Hobbie, J.M. Melillo, B. Moore, B.J. Peterson, G.R. Shaver, and G.M. Woodwell. 1983. Changes in the carbon content of terrestrial biota and soils between 1860 and 1980: A net release of carbon to the atmosphere. Ecological Monographs 53:235-262.

Huston, M., D.L. DeAngelis, and W.M. Post. 1988. New computer models unify ecological theory. BioScience 38:682-691.

Intergovernmental Panel on Climate Change. 1990. Scientific Assessment of Climate Change. Draft report. World Meteorological Organization and United Nations Environment Program.

King, A.W., W.R. Emanuel, and R.V. O'Neill. 1990. Linking mechanistic models of tree physiology with models of forest dynamics: Problems of temporal scale. Forest Growth: Process Modeling of Responses to Environmental Stress. Timber Press, Auburn University, Auburn, Ala.

Lenschow, D.G., and B.B. Hicks (eds.). 1989. Global Tropospheric Chemistry: Chemical Fluxes in the Global Atmosphere. Report of the Workshop on Measurements of Surface Exchange and Flux Divergence of Chemical Species in the Global Atmosphere. National Center for Atmospheric Research, Boulder, Colo. 107 pp.

Logan, J.A., M.J. Prather, S.C. Wofsy, and M.B. McElroy. 1981. Tropospheric chemistry: A global perspective. J. Geophys. Res. 86:7210-7254.

Maier-Reimer, E., and K. Hasselman. 1987. Transport and storage in the ocean—an inorganic ocean-circulation carbon storage cycle model. Climate Dynamics 2:63-90.

Manabe, S., and R.T. Wetherald. 1975. The effects of doubling the CO_2 concentration on the climate of a general circulation model. J. Atmos. Sci. 32:3-15.

Markar, M.S., and R.G. Mein. 1987. Modeling of evapotranspiration from homogeneous soils. Water Resour. Res. 23:2001-2007.

Melillo, J.M., J.D. Aber, and J.F. Muratore. 1982. Nitrogen and lignin control of hardwood leaf litter decomposition dynamics. Ecology 63:621-626.

Moore, B., B. Bolin, A. Bjorkstrom, K. Holmen, and C. Ringo. 1989a. Ocean carbon models and inverse methods. Pp. 409-449 in D.L.T. Anderson and J. Willegrand (eds.), Oceanic Circulation Models: Combining Data and Dynamics. Kluwer Academic Publishers.

Moore, B., M.P. Gildea, C.J. Vorosmarty, D.L. Skole, J.M. Melillo, B.J. Peterson, E.B. Rastetter, and P.A. Steudler. 1989b. Biogeochemical cycles. Pp. 113-141 in M.B. Rambler, L. Margulis, and René Fester (eds.), Global Ecology: Towards a Science of the Biosphere. Academic Press, Inc.

National Research Council. 1984. Global Tropospheric Chemistry. National Academy Press, Washington, D.C.

National Research Council. 1988. Toward an Understanding of Global Change: Initial Priorities for U.S. Contributions to the International Geosphere-Biosphere Program. National Academy Press, Washington, D.C.

Oeschger, H., U. Siegenthaler, and A. Gugelman. 1975. A box diffusion model to study the carbon dioxide exchange in nature. Tellus 27:168-192.

Office for Interdisciplinary Earth Studies. 1985. Opportunities for Research at the Atmosphere/Biosphere Interface. OIES, Boulder, Colo.

Office for Interdisciplinary Earth Studies. 1989. Arctic Interactions. OIES, Boulder, Colo.

Parton, W.J., J.W.B. Stewart, and C.V. Cole. 1988. Dynamics of C, N, P, and S in grassland soils: A model. Biogeochemistry 5:109-131.

Pastor, J., and W.M. Post. 1985. Development of a linked forest productivity-soil carbon and nitrogen model. ORNL/TM-9519. Oak Ridge National Laboratory, Oak Ridge, Tenn. 162 pp.

Pastor, J., and W.M. Post. 1986. Influence of climate, soil moisture, and succession on forest carbon and nitrogen cycles. Biogeochemistry 2:3-27.

Pastor, J., and W.M. Post. 1988. Response of northern forests to CO_2-induced climate change. Nature 344:55-58.

Riley, G.A. 1951. Oxygen, phosphate, and nitrate in the Atlantic Ocean. Bulletin of the Bingham Oceanographic Collection, XIII, Article 1. Yale University, New Haven, Conn.

Rosenzweig, C., and R. Dickinson (eds.). 1986. Climate-Vegetation Interactions. Report OIES-2. Office for Interdisciplinary Earth Studies, Boulder, Colo.

Rosswall, T., R.G. Woodmansee, and P.G. Risser (eds.). 1988. SCOPE 35: Scales and Global Change: Spatial and Temporal Variability in Biospheric and Geospheric Processes. John Wiley and Sons, Chichester, England.

Running, S.W., and J.C. Coughlan. 1988. A general model of forest ecosystem processes for regional applications. I. Hydrological balance, canopy gas exchange, and primary production processes. Ecological Modeling 42:125-154.

Sarmiento, J.L., J.R. Toggweiler, and R. Najjar. 1988. Ocean carbon-cycle dynamics and atmospheric pCO_2. Philos. Trans. R. Soc. London A325:3-21.

Schimel, D.S., M.A. Stillwell, and R.G. Woodmansee. 1985. Biogeochemistry of C, N, and P in a soil catena of the shortgrass steppe. Ecology 66:276-282.

Schimel, D.S., M.O. Andreae, D. Fowler, I.E. Galbally, R.C. Harriss, D. Ojima, H. Rodhe, T. Rosswall, B.H. Svensson, and G.A. Zavarzin. 1989. Priorities for an international research program on trace gas exchange. Pp. 321-331 in M.O. Andreae and D.S. Schimel (eds.), Exchange of Trace Gas Between Terrestrial Ecosystems and the Atmosphere. John Wiley and Sons, Chichester, England.

Schlesinger, M.E. 1983. Simulating CO_2-induced climate change with mathematical climate models: Capabilities, limitations, and prospects. In Proceedings of the Carbon Dioxide Research Conference: Carbon Dioxide, Science and Consensus (CONF-820970), U.S. Department of Energy, Washington, D.C.

Sellers, P.J., Y. Mintz, Y.C. Sud, and A. Dalcher. 1986. A simple biosphere model (SiB) for use within general circulation models. J. Atmos. Sci. 43:505-531.

Shugart, H.H. 1984. A Theory of Forest Dynamics. Springer-Verlag, New York.

Shugart, H.H., and D.C. West. 1980. Forest succession models. BioScience 30:308-313.

Shugart, H.H., M.Y. Antonovsky, P.G. Jarvis, and A.P. Sandford. 1986. CO_2, climatic change and forest ecosystems. Pp. 475-521 in B. Bolin, B.R. Doos, J. Jager, and R. Warrick (eds.), SCOPE 29: The Greenhouse Effect, Climatic Change and Ecosystems. John Wiley and Sons, Chichester, England.

Smith, T.M., H.H. Shugart, D.L. Urban, W.K. Lauenroth, D.P. Coffin, and T.B. Kirchner. 1989. Modeling vegetation across biomes: Forest-grassland transition. Pp. 290-241 in E. Sjogren (ed.), Forests of the World: Diversity and Dynamics (abstracts). Studies in Plant Ecology 18:47-49.

Solomon, A.M. 1986. Transient response of forests to CO_2-induced climate change: Simulation modeling experiments in eastern North America. Oecologia 68:567-579.

Solomon, A.M., M.L. Tharp, D.C. West, G.E. Taylor, J.M. Webb, and J.C. Trimble. 1984. Response of unmanaged forests to CO_2-induced climate change: Available information, initial tests, and data requirements. U.S. Department of Energy, Washington, D.C.

Toggweiler, J.R., K. Dixon, and K. Bryan. 1989. Simulations of radiocarbon in a coarse-resolution world ocean model. 1. Steady state prebomb distributions. J. Geophys. Res. 94:8217-8242.

Vorosmarty, C.V., B. Moore, A.L. Grace, M.P. Gildea, J.M. Melillo, B.J. Peterson, E.B. Rastetter, and P.A. Steudler. 1989. Continental scale models of water balance and fluvial transport: An application to South America. Global Biogeochemical Cycles 3(3):241-265.

World Climate Research Program. 1990. Scientific Plan for GEWEX. May 1990 Draft Report. World Meteorological Organization, Geneva.

Wunsch, C. 1978. The general circulation of the North Atlantic west of 50W degrees determined from inverse methods. Rev. Geophys. Space Phys. 16:583-620.

3
Earth System History and Modeling

OVERVIEW

Contribution of Geologic Studies to Global Change

The geologic record preserves the integrated response of the earth system to a large number of perturbations, including those from human activities, that occurred in the past. Thus the record provides opportunities for comprehensive case studies that can improve our understanding of future changes in the global environment. Furthermore, the geologic record is the only source of information (Table 3.1) on how the climate system has evolved through time. Observations often challenge our constructs of how the earth operates as a system and provide a valuable time perspective for understanding the consequences of future environmental change. Specific important geoscience contributions to global change research include the following:

1. The geologic record provides an independent data set for validating

---

This chapter was prepared by the working group on Earth System History and Modeling established under the Committee on Global Change. Members of the working group were Ellen Mosley-Thompson, Ohio State University, Chair; Eric Barron, Pennsylvania State University; Edward A. Boyle, Massachusetts Institute of Technology; Kevin Burke, National Research Council; Thomas Crowley, ARC Technology; Lisa Graumlich, University of Arizona; George Jacobson, University of Maine; David Rind, Goddard Institute of Space Studies; Glen Shen, University of Washington; and Steve Stanley, Johns Hopkins University. Richard Poore, U.S. Geological Survey, and William Currey, National Science Foundation, participated as liaison representatives from the Committee on Earth Sciences.

TABLE 3.1 Characteristics of Natural Archival Systems (modified from IGBP, 1989, p. 4)

Record	Maximum Temporal Precision	Extent (years)	Derived Parameters[a]
Historical records	Daily	10^3	T P B V M L S
Tree rings	Seasonal	10^4	T P Ca B V M L S
Ice cores			
Polar	1 yr	10^5	T P Ca B V M S
Mid-latitude	1 yr	10^4	T P Ca B V M S
Corals	1 yr	10^5	T Cw L
Pollen and other fossils	100 yr	10^5	T P B
Sedimentary deposits			
Aeolian	100 yr	10^6	T P B V M
Fluvial	1 yr	10^6	P V M L
Lacustrine	1 yr	10^6	T B M
Marine	100 yr	10^7	T Cw B M
Soils	100 yr	10^5	T P B V

[a] Parameters are as follows:

T = temperature
P = precipitation, effective moisture, or humidity
C = chemical composition of air (a) or water (w)
B = vegetation biomass or composition
V = volcanic eruptions
M = magnetic field
L = sea level
S = solar activity

the response of climate models to altered boundary conditions. Such information is critical for assessing the capabilities of models to predict accurately the future consequences of human activities.

2. The geologic record provides valuable information about how different components of the environment are coupled. This information contributes to one of the key goals of the USGCRP.

3. The geologic record is the only available source of information on how the biosphere responds to large changes in the environment. Such information will be particularly valuable for assessing the consequences of future environmental perturbations on the biosphere.

Previous studies demonstrated how the geosciences have contributed significantly to our understanding of how the earth works as a system. For example, the CLIMAP group (CLIMAP Project Members, 1976, 1981) provided the first comprehensive maps of the surface of the earth during the last glacial maximum (18,000 before present (B.P.)). These boundary conditions have been used in a number of modeling experiments, which have provided significant insight on high-latitude climate patterns. Subsequent observational studies by the COHMAP group (COHMAP Members, 1988) extended paleoclimatic mapping efforts to the last deglaciation and the present Holocene interglacial and demonstrated some excellent agreement between models and data for the evolution of the African-Asian monsoon during the early Holocene. Results from the SPECMAP group verified a strong relationship between changes in the earth's orbit and fluctuations of Pleistocene climate (Hays et al., 1976; Imbrie et al., 1984, 1989). Additional studies have shown that natural variations in carbon dioxide and abrupt transitions are an integral part of glacial-interglacial climatic changes (Barnola et al., 1987; Broecker and Denton, 1989; Fairbanks, 1989).

Although significant strides have been made in understanding past environments, there are a number of important problems that require enhanced study. These problems are the targets of the research initiatives outlined in this chapter. In keeping with the philosophy of the USGCRP, both observational and modeling needs are identified for each goal. Before the initiatives are discussed, it should be noted that the emphasis in this report is strongly oriented toward climate, the closely linked environmental changes (e.g., those having to do with oceanic or atmospheric chemistry), and their interactions with the biosphere. Considerations of the role of solid earth processes in global change form the focus of a different element of the program.

Specific Research Initiatives

The research proposed in this chapter addresses three primary topics, each of which falls naturally into a different time scale. An important element of the research program will be to develop global-scale data bases to understand the processes operating within a certain time scale and possible interactions among processes operating on different time scales. The specific initiatives that the committee recommends are as follows:

• to establish an integrated set of globally extensive, high-resolution records of the Holocene (last 10,000 years) as a frame of reference for comparison with any future warming due to greenhouse gases.

• to understand glacial-interglacial fluctuations of the Quaternary. This research will focus on determining how the climate system responded to known forcing (Milankovitch cycles). Such studies will provide valuable information about interactions among components of the climate system, especially biogeochemical cycles and climate. These studies should also enhance our understanding of instabilities in the climate system, as these processes appear to have played a key role in glacial-interglacial climatic changes.

• to examine the system response to large changes in forcing due to carbon dioxide and land-sea distribution changes. This research will provide insight into the nature of warm climates, offer a strong test of climate models, and produce the only available information on the effects of large environmental perturbations on the biosphere.

Priorities

Within these three main areas, the committee recommends that the following topics be given highest priority:

• Holocene high-resolution records for the last 1,000 to 2,000 years;
• Glacial-interglacial cycles, with special emphasis on (a) abrupt system changes and the deglaciation sequence, (b) the carbon cycle, (c) tropical environments at the last glacial maximum, and (d) coupling of different components of the climate system.
• System response to large forcing, with special emphasis on (a) the environment of extreme warm periods such as the Pliocene warm interval (3 to 5 million years ago (Ma)) and (b) evaluation of the climate-biosphere connection during periods of major climatic change such as the Eocene-Oligocene (30 to 40 Ma) transition.

Themes of the Proposed Research

Cutting across all of the proposed topics are some unifying themes: (1) abrupt transitions; (2) climate of warm periods; (3) system response to known forcing; (4) biotic response to climatic change; and (5) coupling between different components of the geosphere and biosphere, with particular emphasis on the carbon cycle. All of these themes represent unique contributions of geologic studies to global change.

Implementation of the Research Plan

Previous research experience demonstrates that progress in understanding past environments has resulted from both individual research projects and more organized efforts such as CLIMAP and COHMAP. Although individual research projects will continue to be an important component of future investigations, it is also apparent that fully successful implementation of some elements of the research plan will require some sustained levels of coordination in which production of large-scale data sets will be necessary.

HOLOCENE HIGH-RESOLUTION ENVIRONMENTAL RECONSTRUCTIONS

During the next few decades a major warming is anticipated in response to the enhanced greenhouse effect. However, there is considerable uncertainty regarding the potential magnitude and regional response to the perturbation. One significant concern is the fact that climate projections do not match the global temperature record of the last century. This disagreement stems in part from the fact that other processes operating in the climate system (e.g., solar forcing, volcanism, and internal variations in the ocean-atmosphere system) may significantly modify temperatures and perhaps mask any greenhouse signal during the early stages of a perturbation.

To clarify the course of future climatic change, it is essential to understand the origin of the natural variability within the environmental system on a time scale ranging from years to centuries. The Holocene (last 10,000 years) record of climatic change offers the temporal and spatial detail necessary to characterize that variability. To date, much of our understanding of Holocene climate is based on a spatially limited data set drawn largely from Western Europe and North America. A broader spatial distribution of historical and proxy records is needed to provide the critical perspective, or backdrop, against which the impact of recent anthropogenic perturbations to the global system can be assessed. Considerably more work is required to develop an adequate understanding of the processes operating on this time scale.

The major focus of this initiative involves determining and understanding decadal- to millennial-scale climate variability by developing a high-resolution global data set. A two-pronged approach is proposed to address this problem: (1) development of a high-resolution global network of climate fluctuations for the last 1,000 to 2,000 years, with special emphasis on the Little Ice Age (LIA) and on process studies for some key regions and (2) development of longer, multiproxy histories devoted to understanding other centennial- and millennial-scale fluctuations in the Holocene.

The Last 1,000 to 2,000 Years

The latest part of the Holocene provides the best opportunity to study such decadal- and centennial-scale processes in more detail because the observational data base is the most extensive and there have been numerous oscillations during this interval (Figure 3.1). However, the processes re-

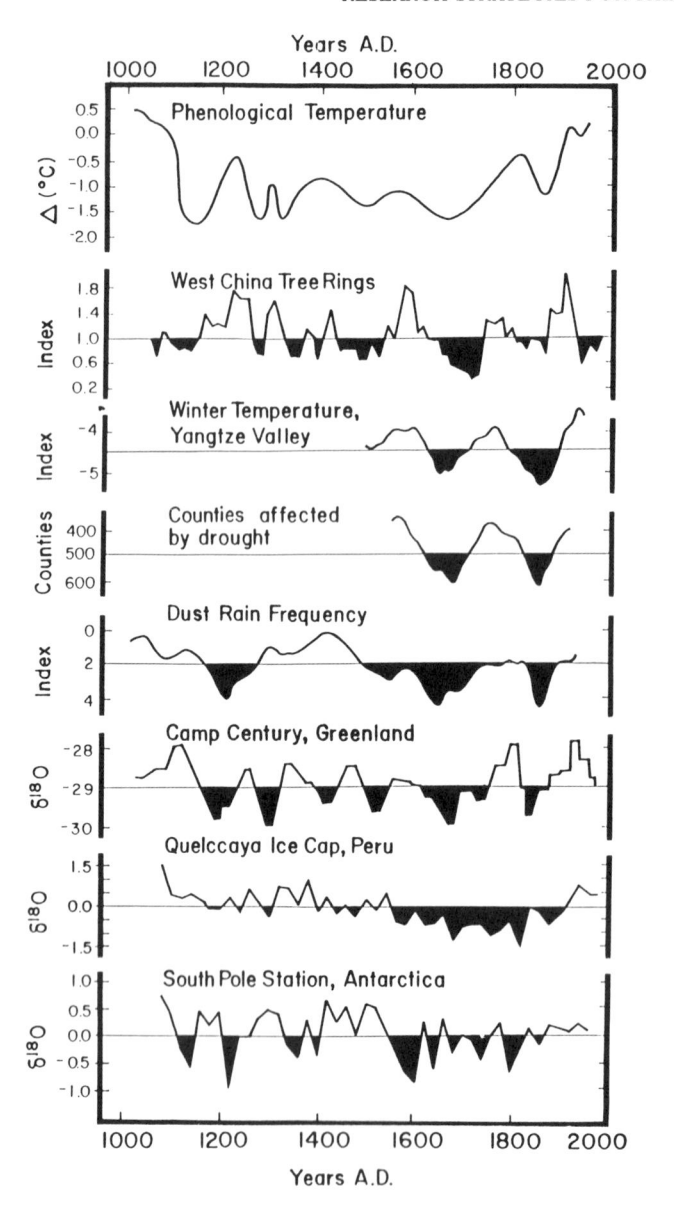

FIGURE 3.1 Evidence for decadal- and centennial-scale oscillations in records spanning the last 1,000 years. Note that oscillations are recorded in several different indices for China (temperature, tree rings, counties affected by drought, and dust rain frequency) and that there is a broad similarity, at least in terms of the time scale of response, with fluctuations in Greenland, Antarctica, and the Peruvian Andes. (Source: Modified from Mosley-Thompson et al., 1990, and Zhang and Crowley, 1989.)

sponsible for these changes are not well understood. This interval is also of particular interest because it encompasses the time of most significant human disturbance of the environment.

Global Network of Environmental Change

Although we have some idea of the frequency and magnitude of climate fluctuations in different regions (Figure 3.1), much less emphasis has been directed toward systematic, detailed comparison of different records to test for synchroneity of change. There are also data gaps in some regions (e.g., parts of the southern hemisphere). The primary goal of this research initiative is to develop a network of 1,000- to 2,000-year records that is sufficiently dense to test for synchroneity of global warming and cooling. Once compiled, these records may be compared with different proposed forcing functions to determine the amount of variance explained by each mechanism.

Observational Needs. Observations are needed

• to determine the timing and spatial variability of past environmental changes on decadal time scales. A global data base of paleoclimate observations is needed. Fortunately, a great diversity of paleoenvironmental sensors are available, many with annual resolution (e.g., direct observations, historical documents, anthropological records, tree rings, ice cores, lake and ocean sediments, and corals). These records must be correlated with an error of less than a decade. An important and nontrivial task will be the development of explicit strategies for combining paleoclimate proxies of varying precision that monitor different elements of the system (e.g., seasonal and geographic sensitivities). Such efforts will require a substantial level of coordination and collaboration.

• to quantify the observed environmental changes in terms of temperature, precipitation, and so on. Preliminary work indicates that cooling during the LIA was on the order of 1.0° to 1.5°C in many places, but it is unclear whether these estimates are mean-annual or seasonal in nature. These efforts should include focused studies, which are essential for understanding the causes and consequences of environmental changes, particularly on regional scales. The latter may be of great significance, as many human activities that are particularly sensitive to environmental perturbations (e.g., food production and transportation) are organized on similar spatial and temporal scales.

• to determine the timing and magnitude of potential changes in forcing. At present the Holocene data base is too sparse to map specific responses at the level of detail necessary to elucidate cause-and-effect relationships. Three likely mechanisms for climatic change on decadal time scales include solar variability, volcanism, and internal nonlinear interactions in the ocean-atmosphere system. Although the timing of solar variability events, based on carbon-14 and beryllium-10 records, is relatively well known (Beer et al., 1988; Stuiver and Braziunas, 1988), the equivalent change in solar forcing is unconstrained. Volcanic fluctuations have also been linked to climatic change (LaMarche and Hirschboeck, 1984). However, one potential record of volcanism (sulfate fluctuations in ice cores) may be complicated by dimethylsulfide (DMS) release in response to changes in oceanic productivity. DMS can be converted into sulfate. It is desirable to acquire methanesulfonic acid (MSA) measurements from ice cores as an independent indication of the ocean productivity component of the ice core sulfate record. Nonlinear interactions in the ocean-atmosphere system may also cause decadal-scale temperature changes (Gaffin et al., 1986; Hansen and Lebedeff, 1978). Testing this idea requires better correlations between changes in the deep ocean (cf. Keigwin and Jones, 1989) and on land. Finally, a quantitative assessment must be made of the amount of variance explained in the climate record by each of these mechanisms.

Modeling Needs. Efforts are required to model the time-dependent variations in temperature as a function of solar variability, volcanism, and ocean-atmosphere coupling. Once observational results allow quantification of the relative magnitude of different forcing agents, various models must be tested to determine if they have the correct sensitivity.

Little Ice Age

A period of special interest is the Little Ice Age (approximately 1450 to 1880 A.D.). In many areas, maximum cooling occurred in the seventeenth century, although not all regions appear to have cooled synchronously. For example, maximum cooling in China may have occurred in the mid-1600s, while in Europe it occurred in the 1690s. During the LIA, there is also evidence for enhanced interannual variability and a stronger meridional circulation. The latter feature may explain some of the regional differences in climate patterns. There are also some indications that transitions into and out of the LIA were relatively abrupt (Thompson and Mosley-Thompson, 1987). Overall, the spatial extent, synchroneity, and magnitude of LIA variations need to be better known.

Observational Needs. Observations are needed

• to develop detailed information about the timing, regional extent, and magnitude of LIA variations, with particular emphasis on the seventeenth century. Although it may not be possible to produce a uniformly dense map of regional climatic change, enough potential information is available from different regions to enhance the synoptic picture of this period. Considerably denser coverage is needed than for the global time series developed for reconstruction of the general patterns of fluctuations over the last 1,000 years discussed above. Information is especially sparse from the tropics and marine areas. In some cases, these voids can be filled by sampling of corals, tropical trees, ice cores, and near-shore or high-resolution marine sediments.

- to better specify the relationship of variability in precipitation and temperature during the LIA. Their trends do not exhibit a simple relationship. In fact, evidence from tree rings (LaMarche, 1974), ice cores (Thompson et al., 1986), and dust records (Zhang and Crowley, 1989) indicates that there were two phases of LIA precipitation, with cool, moist conditions prevailing in the first half and cool, dry conditions dominating the latter half (1700 to 1880 A.D.).
- to investigate apparent abrupt transitions into and out of the LIA. These studies may lead to identification of potentially important, but less obvious, causal mechanisms operating on shorter time scales within the Holocene. These may arise from changes in transient geochemical reservoirs (e.g., ice and labile carbon stores), strong feedbacks (e.g., albedo and carbon dioxide), or volcanism (Berger and Labeyrie, 1987). The Holocene record is rich in evidence for rapid climate changes in many regions, and a systematic search for widely correlated events reflecting large-scale, short-term climate shifts is recommended.

Modeling Needs. Models are needed to construct and test three-dimensional circulation models of the atmosphere and oceans relating specific forcing mechanisms and known system responses (i.e., observations). Model results can be compared to the inferred response in regions where records are available. These experiments will prove valuable for determining the sensitivity of models (and the real world) to known forcing. The abundant high-resolution data for the LIA are particularly appropriate for investigating potential forcings and the climatic and biospheric responses.

Regional Process Studies

Knowledge of the processes responsible for local changes inferred from proxy records is essential for accurate interpretation. Often this information provides additional insight into regional processes that strongly affect both local- and global-scale circulation systems. Therefore regional climatic chronologies should be developed for areas where episodic regional-scale processes strongly affect both the local climate and global-scale circulation systems. Three candidate areas that should be considered for further study are (1) the subpolar North Atlantic basin, (2) the equatorial Pacific basin, and (3) the Asian monsoon.

Observational Needs. Observations are required

- to develop a greater understanding of processes occurring in the subpolar North Atlantic basin. Geologic studies suggest that this region may be a key area for understanding possible changes in the oceanic-atmospheric circulation (see the section "The Last 40,000 Years" below). This area encompasses one of the densest arrays of historical data, and thus it is desirable to determine the pattern of climate fluctuations in this region on decadal scales and to ascertain whether they were accompanied by any changes in the oceanic circulation. A coordinated effort linking the climate of eastern North America, Greenland, Western Europe, and the subpolar North Atlantic is recommended. This effort will require the acquisition of very high sedimentation rate deep-sea records (see section "Sample Acquisition" below) from shallow marine areas or sediment drifts for evaluation of possible changes in the surface and deep circulation.
- to develop long time series of El Niño-Southern Oscillation (ENSO) fluctuations. In the past decade, researchers have demonstrated the large-scale nature of ENSO events and their very important influence on tropical rainfall patterns. Ice cores and corals contain information about interannual climate variability and offer the opportunity to extend these records back several centuries or more. Such results could provide an enhanced understanding of ENSO. This research will require additional information on tropical rainfall from tree rings, ice cores, and upwelling variations as recorded in coral reefs.
- to develop long time series of monsoon fluctuations. The Asian monsoon is one of the most important features of the planetary circulation, and fluctuations in its intensity affect the lives of nearly 2 billion people. Long time series are available from India extending back about 100 years, and historical time series from China extend back at least 500 years (Zhang and Crowley, 1989). However, more information is needed to understand the temporal variations.

Modeling Needs. Models need to be developed

- to simulate many of the features of observed oceanic-atmospheric anomalies such as ENSO events (Cane et al., 1986). Using existing models, the sensitivity of such regional processes to observed or suspected changes in other components of the system can be examined. Conversely, if observations clearly indicate a change in frequency or character of oceanic-atmospheric anomalies, models may suggest potential causes. These efforts will contribute to validating the utility of such models for predicting possible future changes in ENSO-type events due to global warming.

- to link climate fluctuations on decadal- and centennial-time scales with the present generation of atmospheric and oceanic models used to study the above regional processes. At present the models can theoretically generate variance on decadal and longer time scales, but key parameterizations in the models are not well constrained by observations. Approaching the problem from both an observational and a modeling viewpoint for the last 500 years may provide additional insight to processes occurring on shorter time scales.

Earlier Holocene Millennial-Scale Fluctuations

Geologic records indicate that LIA-type fluctuations occur on a characteristic time scale of 2,000 to 3,000 years over much of the last 20,000 years (Figure 3.2). Therefore, any explanation for climatic variability in the last several thousand years should be applicable to these earlier fluctuations.

Observational Needs. Observations are needed to develop time series of system response from selected regions for the last 10,000 years. Information is available from such areas as mountain glaciers, the central Asian highlands, and African lakes (Röthlisberger, 1986; Street-Perrott and Harrison, 1984; Thompson et al., 1989). Some additional high-resolution marine records are critically needed. High-resolution records in the North Atlantic might provide information about fluctuations of the ocean on this time scale. Although records of solar variability extend back to 9,600 B.P. (Stuiver and Braziunas, 1988), the record of volcanism in both hemispheres is not as well documented. High-resolution terrestrial records, especially ice cores, should contribute substantially to reconstruction of the earth's volcanic history during the Holocene.

Modeling Needs. Efforts are needed to test models of climate variability developed for the last 1,000 years against longer records. Any explanation for decadal- to millennial-scale fluctuations of the last 1,000 years should also be applicable to earlier time intervals. Specific models should test this hypothesis.

GLACIAL-INTERGLACIAL CYCLES

The U.S. Global Change Research Program seeks to improve our understanding and predictive capabilities of the climate system's response to

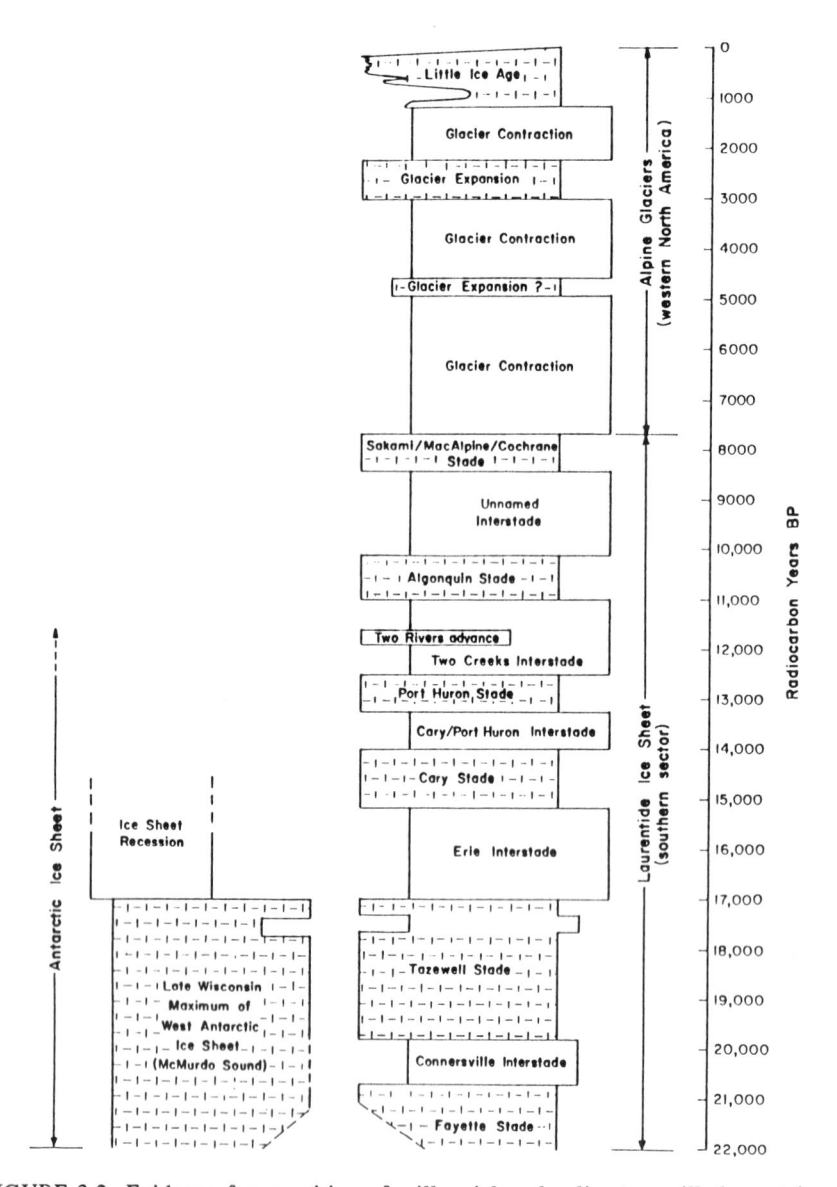

FIGURE 3.2 Evidence for repetition of millennial-scale climate oscillations at intervals of 2,000 to 3,000 years over the last 20,000 years. Analysis of any specific event, such as the Little Ice Age or Younger Dryas, must take into account the characteristic time scale of these oscillations. (Reprinted, by permission, from P.A. Mayewski et al. (1981). Copyright © 1981 by John Wiley and Sons, Inc.)

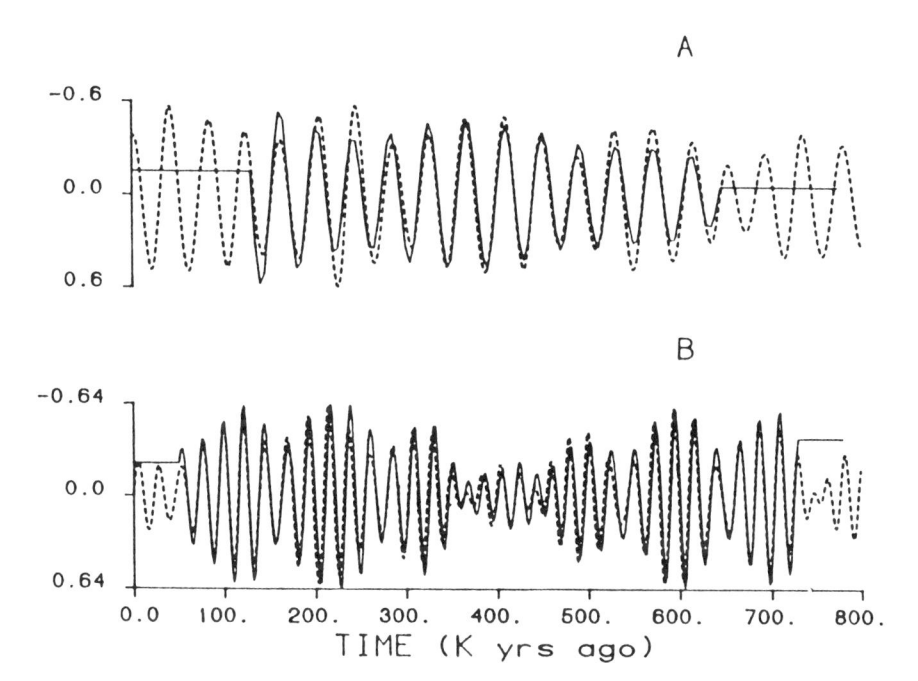

FIGURE 3.3 Filtered oxygen isotope record for the late Pleistocene, illustrating very high coherence between orbital forcing (dotted line) at the (a) obliquity and (b) precession bands (40,000- and 23,000-year periods, respectively) and global ice volume response (solid line). (Reprinted from J. Imbrie et al. (1984). Copyright © 1984)

known forcing. Climate records of the late Pleistocene (Figure 3.3) demonstrate significant responses to orbitally induced variations in insolation—the Milankovitch effect (Imbrie et al., 1984). Carbon dioxide and methane in the atmosphere also vary with glacial-interglacial cycles. The geologic record indicates that the system response is quite complex in space and time and that there are probably instabilities in the climate system.

The principal objective of this initiative is to determine the nature of climate system responses to known forcing during the Pleistocene. These studies will provide information about the characteristics of the coupling among different components of the climate system and therefore enable development and testing of models describing these interactions. These studies also will be especially valuable for investigation of climate instabilities and biogeochemical cycles.

To meet the objectives of this initiative, the research strategy must be divided into different subtasks. These subtasks fall naturally into different time scales, with processes operating on one time scale sometimes affecting processes operating on different time scales. The subtasks involve detailed studies over the range of the carbon-14 time scale (see the section "The Last 40,000 Years" below), climate fluctuations over the last glacial cycle (see the section "The Last Glacial Cycle (Last 130,000 Years)" below), and climate fluctuations over several glacial cycles (see the section "The Last Few Glacial-Interglacial Cycles (Last 500,000 Years)" below). Accomplishing these goals requires both field programs for data collection and modeling studies.

The Last 40,000 Years

Numerous studies demonstrate that there have been several glacial cycles of approximately 100,000-year duration during the late Pleistocene (Imbrie et al., 1984). High-resolution time series of climate variables record a significant response to orbital variations over this interval. However, the exact manner in which the orbital signal is transmitted through the climate system is not understood. The interactions between external forcing and system response involve both gradual processes (COHMAP Members, 1988) and abrupt transitions that may reflect instabilities in the climate system (e.g., Broecker and Denton, 1989).

Much progress has been made in mapping and understanding the evolution of climate over the last 40,000 years (e.g., COHMAP Members, 1988; Crowley and North, 1990). However, a number of important problems remain that require enhanced study to increase our understanding of the climate system and its response to altered boundary conditions. The committee recommends the following research topics for special emphasis: (1) analyses of the abrupt changes that occurred during deglaciation (14,000 to 10,000 B.P.) and possibly from 40,000 to 30,000 B.P. and (2) resolution of model-data discrepancies such as the characterization of tropical environments at the last glacial maximum and the environment of mid-continental Eurasia at 6,000 B.P. The observational needs and modeling requirements are discussed for each.

Abrupt Changes

Some of the most spectacular examples of abrupt climatic change in the earth's history are known to have occurred at the end of the last glacial stage (14,000 to 10,000 B.P.). For at least some regions of the world, there is growing evidence for an abrupt (1,000 years) warming at 14,000 B.P., followed by a cooling at about 11,000 B.P. (termed the Younger Dryas) and another abrupt warming about 10,000 B.P. (Figure 3.4). New sea level estimates (Fairbanks, 1989) link meltwater pulses to these warming events, revealing that sea level rose 24 m in less than 1,000 years at 12,000 B.P.

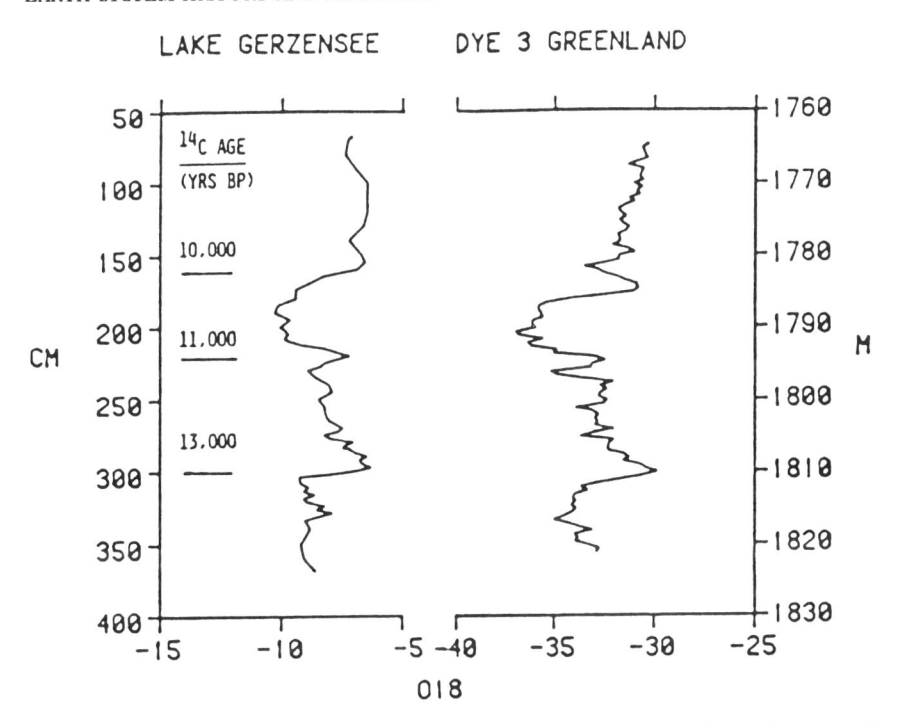

FIGURE 3.4 Evidence from Greenland and Switzerland for rapid environmental oscillations at the end of the last glacial. Note the striking similarity of warming about 14,000 B.P., cooling at 11,000 B.P., and warming again at 10,000 B.P. (From Oeschger (1985). Copyright © 1985 by the American Geophysical Union.)

There is also evidence for abrupt changes in surface, intermediate, and deep water during the Younger Dryas (Boyle and Keigwin, 1987; Ruddiman and McIntyre, 1981). Some studies suggest the 10,000 B.P. transition to warm conditions may have occurred in as little as 20 years (Dansgaard et al., 1989). Abrupt transitions found in the Dye 3 Greenland ice core between 40,000 and 30,000 B.P. may be associated with rapid carbon dioxide fluctuations (Oeschger et al., 1985). However, these earlier events have yet to be reproduced in other ice cores. Analyses of pollen data have shown that the rate of change in terrestrial vegetation has varied considerably through time, with widespread and abrupt changes in eastern North America concentrated especially in the period from 14,000 to 10,000 years ago, probably in response to large-scale changes in atmospheric circulation (Jacobson et al., 1987).

The abruptness and magnitude of the above changes remain unexplained. They represent a dramatic contribution of earth studies to the understanding of global change. Some ideas link the rapid changes to a complete reorganization of the ocean-atmosphere system (Broecker and Denton, 1989; Broecker et al., 1985). To better understand this phenomenon, it is necessary to determine the relationship between slowly changing boundary conditions (e.g., Milankovitch forcing; COHMAP Members, 1988) and abrupt system responses.

Observational Needs. Observations are required

• to determine the magnitude and global extent of abrupt changes during the last deglaciation (14,000 to 10,000 B.P.). Previous studies reveal large coherent changes in the North Atlantic basin, with changes at 14,000 to 13,000 B.P. also occurring in the southern hemisphere. However, the chronology of these events must be improved, both on land and sea, and we need better measurements of the extent and magnitude of the system response. Needed are multiple, independent monitors of changes on land (i.e., temperature, precipitation, and biota), in the surface and deep ocean, and in the atmosphere (i.e., dust and sulfate aerosols) including its chemical composition (carbon dioxide and methane). Also needed is an even more intensive study of climatic and biospheric changes in the North Atlantic basin, as this region seems to be a key to understanding the processes responsible for the changes. A precise and much-improved chronology of events in key parts of the ocean-atmosphere-biosphere system can be developed now by application of atomic mass spectroscopy (AMS) carbon-14 techniques, which have the advantage of requiring much smaller and more reliable samples than earlier techniques. These results should be viewed within the framework of global-scale data bases as mapped by the COHMAP Members (1988).

• to test the existence of rapid carbon dioxide fluctuations from 40,000 to 30,000 B.P. This subtask is closely related to a more comprehensive examination of carbon cycle fluctuations, which is best accomplished when viewed from the perspective of the last 130,000 years (see the section "Global Carbon Cycle" below). Unlike the deglaciation, in which carbon dioxide changes apparently were not rapid, studies of the Dye 3 Greenland ice core reveal rapid (centennial- or even decadal-scale) oscillations in carbon dioxide and other climate variables (Dansgaard et al., 1989). Such discoveries have prompted hypotheses that the ocean-atmosphere system may have more than one stable state (Broecker et al., 1985). Those rapid fluctuations have not been detected in the Byrd antarctic ice core (Neftel et al., 1988). Possible explanations for this contradiction include (1) the Greenland site was exposed to local warming, with seasonal melting affecting carbon dioxide concentrations or (2) the air "closure time" for the Byrd core is too long. To clarify these issues, it is highly desirable to gather ice core records from regions of higher accumulation where air closure times are shorter and time scales can be established with greater confidence. Any evidence linking

EARTH SYSTEM HISTORY AND MODELING 83

rapid changes in both climate and carbon dioxide is important for validating predictions for future climate. The second International Greenland Ice Sheet Program (the U.S. program is GISP II and the European program is GRIP) has just begun drilling in central Greenland to obtain a 150,000-year record with moderate time resolution. Studies of gas bubbles, isotopes, major element chemistry, and dust have already begun; other measurements should be added to take advantage of this core. In the future it may be necessary to acquire other ice cores (e.g., from Antarctica).

• to determine the nature of biotic responses to abrupt environmental change. Although major terrestrial extinctions occurred at the end of the Pleistocene, many large environmental changes were not associated with extinctions. The background of gradual biotic responses to orbital forcing has been well illuminated by recent synoptic studies (e.g., COHMAP Members, 1988; Huntley and Webb, 1989; Huntley and Prentice, 1988; Webb et al., 1987). There are, nevertheless, events such as the Younger Dryas cooling (ca. 11,000 to 10,000 B.P.) associated with large, rapid changes in some regions of the world. Understanding of the spatial extent and synchroneity of biotic responses to both abrupt and gradual climatic changes must be greatly improved.

• to determine the temporal history of millennial-scale fluctuations that may be linked to abrupt transitions. Some data indicate that abrupt transitions such as the Younger Dryas and possible oscillations between 40,000 and 30,000 B.P. may be related to 2,500-year time scale climate oscillations that have been detected in the Holocene (Denton and Karlén, 1973; Figure 3.2), including the LIA, and in some marine records (e.g., Pisias et al., 1973). At present, studies of such fluctuations are limited by the lack of long time series from high-deposition-rate deep-sea cores. Enhanced coring capabilities in deep-sea sediments may be needed (see the section "Sample Acquisition" below). On land, expanded paleoecological and ice core studies with high temporal resolution for the past 40,000 years are needed.

Modeling Needs. Models are needed

• to develop a better understanding of the causes and processes involved in rapid climate transitions. With the reality of abrupt transitions becoming more and more apparent, a much better understanding of the causes and processes involved in the transitions is needed. Broecker et al. (1985) have proposed that such changes may be caused by major reorganizations of the ocean-atmosphere system. Some modeling studies support elements of this conjecture (e.g., Maier-Reimer and Mikolajewicz, 1989; Manabe and Stouffer, 1988), and detailed modeling comparisons for the Younger Dryas (11,000 to 10,000 B.P.) support the consistency between ocean and land data in the vicinity of the North Atlantic (Rind et al., 1986). However, future work must apply improved oceanic models, assess the global distribution of the changes, and determine how abrupt transitions are related to slowly changing boundary conditions (Milankovitch forcing).

Resolving Model-Data Discrepancies over the Last 20,000 Years

Detailed global mapping programs of the last decade have provided substantial information about the surface of the Earth over the last 18,000 years (CLIMAP Project Members, 1976, 1981; COHMAP Members, 1988). These studies have stimulated a number of modeling endeavors, which demonstrate levels of agreement between models and data varying from excellent to poor (cf. summary in Crowley and North, 1990). In order to establish higher levels of confidence in climate models, it is necessary to reconcile these differences over a time interval that is particularly rich in data. Better understanding of these "snapshot" time intervals may reveal how slowly changing boundary conditions could trigger instabilities in the climate system (see the section "Abrupt Changes" above).

Some special areas of enhanced model-data comparison involve the following: (1) resolving tropical sea surface temperature (SST), lowland precipitation, and snowline fluctuations at 18,000 B.P. (Rind and Peteet, 1985; Webster and Streten, 1978); (2) making more quantitative comparisons of models and observations for the African-Asian monsoon regions at 9,000 and 6,000 B.P. (COHMAP Members, 1988; Mitchell et al., 1988); (3) clarifying factors responsible for high-latitude climatic change in the southern hemisphere at 18,000 B.P. (cf. Crowley and North, 1990); and (4) utilizing oceanic GCMs to understand intermediate- and deep-water circulation changes during the last glacial maximum and deglaciation (Boyle and Keigwin, 1987; Maier-Reimer and Mikolajewicz, 1989). The committee recommends that resolving model-data discrepancies in the tropics at 18,000 B.P. be given the highest priority.

Observational Needs. Observations are needed

- to increase information on regional climate patterns over the last 20,000 years. Although we have good information from some regions, a number of data gaps must be filled. Some of the most important areas of concern are tropical lowlands and SSTs at 18,000 B.P. The tropical lowland information is especially important for better estimates of temperature and precipitation changes at low elevations on the continents. This information would also be extremely valuable for assessing the sensitivity of tropical rain forests to environmental change. These studies should improve our understanding of high biological diversity in the tropics.
- to improve estimates of temperature and rainfall variations in regions where model-data comparisons need to be more quantitative, such as the

lowland tropics and southwestern United States (18,000 B.P.) and the African-Asian monsoon (9,000 to 6,000 B.P.).

• to develop a more complete assessment of transfer functions used for paleoenvironmental estimates. Paleo-oceanographic data should be carefully evaluated and compared with other, independent estimates of SST. Temperature depression associated with snowline and paleovegetational descent on tropical mountains should be quantified and dated. Estimates for changes in low-elevation temperatures need substantial improvement. Any new paleoecological records from low latitudes for 18,000 B.P. and before will be extremely valuable.

Modeling Needs. Models should be used for

• increased testing of atmospheric and oceanic models in order to resolve apparent model-data discrepancies. New GCMs with better horizontal and vertical resolution are needed. Precipitation estimates derived from models (with biospheric feedback) or geological data should be included. Finally, oceanic GCMs should be used to explore additional changes over the last 18,000 years.

• application of advanced models to address important model-data discrepancies over the last 20,000 years. Among the most important modeling problems to address are (1) capability of climate models to generate drier conditions in tropical lowlands at 18,000 B.P.; (2) reconciliation of apparently small lowland temperature changes with larger upland temperature changes in the tropics at 18,000 B.P.; (3) improved quantitative agreement between models of enhanced monsoon and southwestern U.S. rainfall fluctuations at approximately 9,000 and 18,000 B.P., respectively, and observations; (4) identification of the mechanisms responsible for significant cooling in the high latitudes of the southern hemisphere at 18,000 B.P.; and (5) ability of oceanic models to generate decreased deep-water and enhanced intermediate-water production rates in the North Atlantic at 18,000 B.P.

The Last Glacial Cycle (Last 130,000 Years)

Many of the processes that can be studied over the carbon-14 time scale provoke explanations that should be applicable to other time intervals of climatic change during the Quaternary. In particular, a considerable amount of information is available from marine and ice core records revealing the climate of the last full glacial cycle (Figure 3.5). The growing number of land records available from this time interval will make it possible to map climate evolution over a full glacial cycle as it relates to Milankovitch forcing. Figure 3.5 suggests that carbon dioxide may have played an important role in climatic change over the last 130,000 years. However, because carbon dioxide lags climatic change in the southern hemisphere at the end

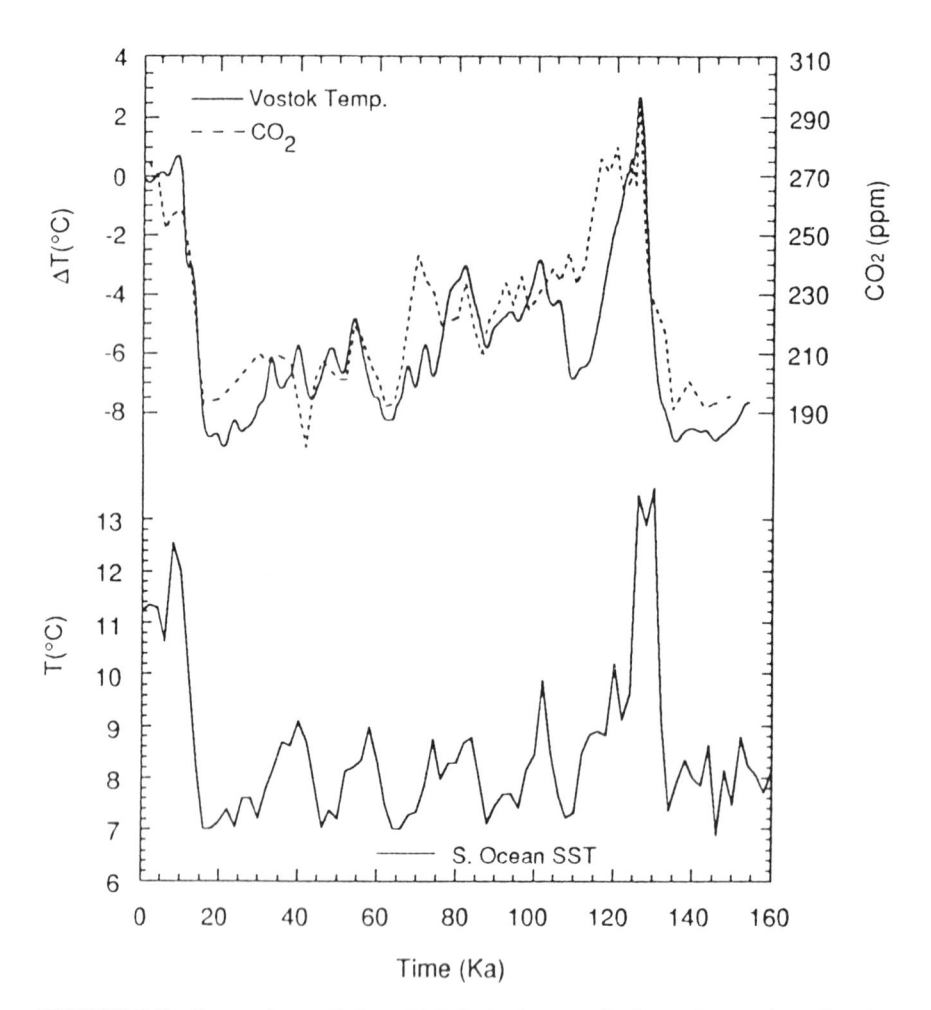

FIGURE 3.5 Comparison of three high-latitude records from the southern hemisphere showing the overall good agreement between carbon dioxide and temperature changes (inferred from ΔD). However, the carbon dioxide record clearly lags the Vostok (Antarctica) temperature record at the end of the last interglacial. (Reprinted, by permission, from T.J. Crowley and G.R. North (1990). Copyright © 1990 by Oxford University Press. Data sources: the Vostok ΔD record (Jouzel et al., 1987) and the carbon dioxide record (Barnola et al., 1987) are plotted according to the revised chronology of Petit et al. (1990).)

of the last interglacial, the climate feedback of this important variable warrants further evaluation. Thus study of climatic change over the last 130,000 years will make it possible to (1) address more fully the mechanisms responsible for ice age carbon dioxide changes and the magnitude of the carbon-dioxide-climate feedback; (2) investigate both the nature of warmth during the last interglacial and the processes responsible for cooling and ice cap growth at the end of the warm period; and (3) trace the regional variations in climate on both land and sea as the system evolves through an entire glacial cycle.

Global Carbon Cycle

The ability to study changes in past atmospheric composition from the bubbles trapped in polar ice caps has been one of the major scientific developments of the past decade. Convincing evidence now exists for changes in atmospheric carbon dioxide during the past 150,000 years (Barnola et al., 1987; Neftel et al., 1988). Ongoing work is documenting variations in methane, nitrous oxide, and the oxygen isotope composition of ancient atmospheres; the data show that large changes in atmospheric carbon dioxide have occurred that are approximately in step with the major climatic changes of the last 150,000 years and that significant changes have occurred in other natural greenhouse gases. One particularly intriguing result is the observation that the antarctic climate appears to have cooled substantially before carbon dioxide decreased at the end of the last interglacial period (Figure 3.5). This observation needs to be confirmed and examined for its climatic implications.

The causes of changes in atmospheric carbon dioxide must be sought in models of the oceanic carbon system, which is the only buffer sufficiently massive, yet fast enough to drive the large and rapid changes seen in the ice core record. Several competing ideas remain to be tested concerning atmospheric carbon dioxide (e.g., Boyle, 1988; Broecker, 1982; Knox and McElroy, 1984; Sarmiento and Toggweiler, 1984; Siegenthaler and Wenk, 1984), and no consensus exists as to which, if any, of these models provides the correct explanation for the observed atmospheric changes. The continued development of oceanic carbon system models to explain past changes in atmospheric carbon dioxide should be a high priority in global change research.

Observational Needs. Observations are needed

to develop a better understanding of the timing of events in the ocean during the last 150,000 years. This information is needed for evaluation of scenarios for climatic change and for measurement of time constants associated with components of the climate system. As mentioned in the preceding section "Abrupt Changes," there have been major changes in both the deep- and the intermediate-water circulations, which have implications for atmospheric carbon dioxide. While the ocean must be the proximate determinant of changes in atmospheric carbon dioxide, it is difficult to evaluate precisely the relative timing of events observed in the ocean cores and ice cores. Furthermore, it is also necessary to understand the temporal relationships among changes in ocean circulation, ocean chemistry, and changes on the continent (e.g., in ice extent, atmospheric dust transport, and vegetation). As another example, we need to know the timing of glacial mass-wasting in various parts of the northern hemisphere ice sheets in relation to changes in oceanic temperature and salinity; can the sequence of events in the ocean be linked to forcing by meltwater and iceberg calving?

- to conduct detailed comparisons of ice core, marine, and continental records. Some problems of particular interest that might be examined include the interactions of terrestrial ecosystems, environmental change, and atmospheric composition, and the climatic effect of possible variations in cloud cover. The latter might be induced by changes in atmospheric dust measured in both ice cores and deep-sea cores and by changes in DMS emissions, which have been inferred from ice core measurements of the DMS by-products (MSA and excess sulfate). Both of these variables could affect the radiation budget on glacial-interglacial time scales (Charlson et al., 1987; Harvey, 1988).

Modeling Needs. Models are required

- to understand the origin of ice age carbon dioxide and methane changes. One of the key modeling challenges facing earth scientists is to understand the origin of the ice age carbon dioxide fluctuations. Attempts to elucidate this relationship over the past decade have turned up a surprising number of ways in which the ocean might change the partial pressure of carbon dioxide (pCO_2). Yet this research has been frustrated by difficulties in accounting for the timing and amplitude of the pCO_2 observations. Current understanding of the role of the ocean circulation, chemistry, and productivity does not account for the observed pCO_2 record. An effort should be made to improve models of oceanic carbon dioxide. Such efforts will be a major contribution of earth system research to the USGCRP. There is a growing need to couple climatic and geochemical models in the future.

The Previous Interglacial

The previous interglacial was the last time when conditions were as warm as the present (Holocene) interglacial. Although a number of studies have concluded that it was warmer than at present, other investigations suggest that in some cases the warmth was primarily seasonal in nature (e.g., Prell and Kutzbach, 1987). Except for a few regions, global SSTs may not have been significantly different from those at present (CLIMAP Project Members, 1984). However, sea level was about 5 to 6 m higher than at present (e.g., Mesolella et al., 1969; Dodge et al., 1983).

It has been suggested that the last interglacial may be an "analog" for the early stages of a future greenhouse warming (e.g., Hansen and Lebedeff, 1987). Acceptance of this suggestion, however, requires clarification of the seasonal (versus year-round) nature of the warmth and whether the warmth was globally synchronous. Present evidence suggests that during the last interglacial globally averaged mean annual temperatures were not significantly greater than at present (Crowley, 1990) and that the last interglacial should not be cited as a carbon dioxide analog. Nevertheless, it is still desirable to understand the regional patterns of warmth during this period. It is also important to determine which ice sheets contributed to the sea level rise—both Greenland and the West Antarctic ice sheets have been suggested (Koerner, 1989; Mercer, 1978). These two topics—warmth and sea level—are critically related because knowledge of the magnitude of warming may help calibrate the sensitivity of the cryosphere to warming trends and thus enhance our ability to predict the course of future changes in sea level.

Additional questions related to this time period concern the transition into the last glacial stage. As stated earlier, carbon dioxide lags cooling in the southern hemisphere (Figure 3.5). Furthermore, recent modeling studies have suggested that the reduced solar insolation of the 115,000 to 105,000 B.P. interval may not have been sufficient to generate or maintain low-elevation ice sheets in the Laurentide area, even with reduced carbon dioxide (Rind et al., 1989). Such results call into question either climate model sensitivity or our understanding of the mechanisms whereby orbital variations lead to ice sheet growth, or both.

Observational Needs. Observations are required

- to determine the magnitude and nature of the last interglacial warmth. Two items of paramount importance involve (1) evaluation of whether times of greater warmth were globally synchronous and (2) evaluation of terrestrial paleoclimate proxy-data to determine whether the warmth was seasonal or mean-annual in nature. The latter may require new techniques for estimating past temperatures.
- to establish the source of global rise in sea level. The higher temperatures presumably triggered a sea level increase of 5 to 6 m. However, we need to know whether the increase resulted from melting in Greenland or in Antarctica, or both.
- to clarify the nature of climate transitions into and out of the last interglacial. Research on the last deglaciation suggests that significant cli-

mate oscillations were involved—perhaps similar to the Younger Dryas. Are such oscillations characteristic of all deglaciations? Other studies of relevance include the rate at which climate deteriorated at the end of the last interglacial. Some studies suggest that the transition was quite abrupt (Frenzel and Bludau, 1987). Answering this question involves knowing more about where and when the Laurentide Ice Sheet developed.

Modeling Needs. Efforts are needed

• to develop cryosphere models that incorporate inferred temperature history with fluctuations of ice sheets on Greenland and Antarctica.

• to utilize ocean models to determine how changes in atmospheric forcing affected oceanic circulation. Of particular interest is the question whether such oceanic changes could account for inferences of increased mean-annual temperatures in some regions.

• to continue atmospheric modeling in order to test sensitivity of models to known variations in orbital forcing, in particular with respect to regions of ice sheet growth and decay. It is of special interest to assess the role of carbon dioxide in the onset of glaciation.

Regional Variations in Climate over a Glacial Cycle

Although we have a reasonable picture of how climate changed in a number of regions over the last glacial cycle, changes in different regions have not been adequately integrated to determine the nature of dynamical linkages. The fairly widespread availability or potential availability of a number of good records over this time interval justifies a period of time to be studied as a "special observing period," for which the various mechanisms proposed to explain observed changes may be tested.

Observational Needs. Observations are required

• to develop long terrestrial records of climatic change. At present knowledge of the land record beyond the range of carbon-14 dating is rather limited. In Europe, studies of several exceptionally long stratigraphic records have revealed the responses of vegetation to climatic changes throughout the entire last glacial-interglacial cycle (e.g., Guiot et al., 1989). More records should be acquired and correlated with the deep-sea record. Acquisition of long terrestrial records may require enhanced coring capabilities (see the section "Sample Acquisition" below).

• to refine estimates of variations in dust and clouds. Evidence suggests that these changes may affect the planetary radiation budget and contribute to glacial-interglacial climatic change. The dust record is of large geographic scale (Petit et al., 1990). Clouds, as inferred from DMS by-products in ice cores (Legrand et al., 1988), must be better understood.

The Last Few Glacial-Interglacial Cycles (Last 500,000 Years)

Although detailed investigation of the last glacial cycle will provide considerable insight into processes responsible for Pleistocene ice sheet fluctuations, that information alone will not resolve the problem. These ideas must be tested through several realizations of glacial-interglacial cycles. In addition to the study of the last glacial cycle, the second major proposed task will address selected elements of these longer time series so as to place phenomena from the most recent cycle in perspective.

Previous work by the SPECMAP group indicates a strong influence of orbital forcing on the evolution of the earth's climate system (e.g., Berger et al., 1984). This work represents an excellent opportunity to examine system sensitivity to known forcing. In addition, sampling climate fluctuations through several realizations enables gathering of reliable statistics indicating how various components of the system are coupled. Such results can make valuable contributions to the USGCRP. For example, studies clearly indicate that there are significant phase offsets among the different components of the climate system (e.g., Imbrie et al., 1989). Although significant progress has already been made on this topic, better information about some variables is needed to constrain models of the Pleistocene ice ages.

Observational Needs. Observations are required

• to continue measurements of various components of the climate system over the last few hundred thousand years. More measurements of SSTs, deep and intermediate waters, aeolian fluxes, and various components of the carbon cycle are needed. It is especially desirable to link these records with the growing number of long land records (e.g., Hovan et al., 1989; Kukla, 1989).

Modeling Needs. Efforts are needed

• to develop time-dependent models linking various components of the climate system. For example, GCM studies indicate that there is a direct link between seasonal variations in orbital forcing and monsoon variability over the last glacial cycle (Kutzbach and Street-Perrott, 1985; Prell and Kutzbach, 1987). Observational studies indicate that this monsoon signature is detectable in the deep sea (Pokras and Mix, 1987; Prell, 1984) and may be involved in continental biomass variations (Keigwin and Boyle, 1985) that could affect atmospheric methane (wetlands are an important source of methane). An ocean modeling study supports a direct link between orbitally induced changes in the monsoon and surface circulation in the Indian Ocean (Luther et al., 1990). More systematic information about these interactions will clarify an important climate problem. Because the relationship between forcing and initial system response is so clear and amenable to modeling, the new information will provide an ideal opportunity for some advanced modeling studies using, for example, biospheric models and coupled oceanic-atmospheric models.

- to develop models of ice age carbon dioxide fluctuations for the last glaciation and glacial cycle that make predictions of the time-dependent history of various components of the climate cycle. Modeling studies need to cast such predictions in the time and frequency domains.

SYSTEM RESPONSES TO LARGE CHANGES IN FORCING

Prior to the Pleistocene, there were large changes in boundary conditions for the earth's climate system. These changes involved variations in continental position and height, the boundaries of the ocean basins, sea level, and probably the carbon dioxide content of the atmosphere. The evolving boundary conditions were paralleled by large changes in the earth's climate. The major features of the evolution of global climate over the last 100 million years are illustrated in Figure 3.6. Note the long-term cooling trend characterized by relatively abrupt transitions. Although these long-term trends are based on qualitative and semiquantitative information in both marine and continental records, more quantitative information is now becoming available.

The environment of the Cenozoic offers unique opportunities for studies in global change. Topics of special importance involve the response of the earth system during times when climates were substantially warmer than modern and intervals during which the climate system experienced rapid and large changes. An especially important contribution involves the biospheric response to these large changes, as the geologic record provides the only information available on the relation between extinction events and environmental change.

Pre-Pleistocene climate records also present a formidable challenge to models. Detailed information on past warm intervals will be extremely valuable for testing model ability to simulate warmer climatic conditions that are likely to have resulted for different reasons. As these same models will be used to predict the nature of future greenhouse warming, it is important to validate the models in an independent manner. The geologic record provides the only independent test of these models. Cenozoic records will be especially useful in testing oceanic models and coupled oceanic-atmospheric models. It is especially important to test the highly parameterized coupling between the ocean and atmosphere against independent data sets because the boundary conditions are so radically different. There is a wealth of marine data available from the Ocean Drilling Program (ODP) to validate model simulations. Important questions to consider include whether models are capable of simulating warmer climates under different boundary condi-

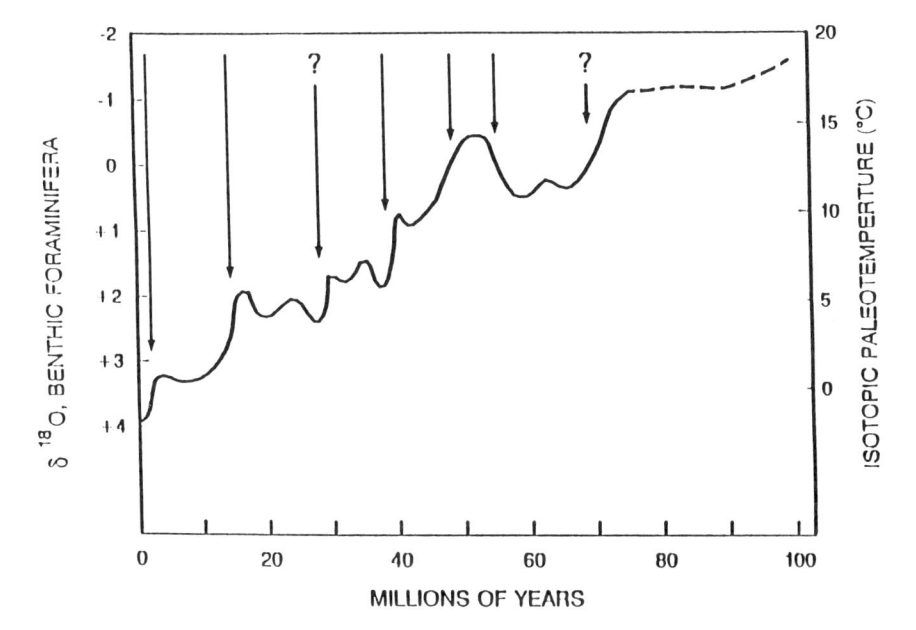

FIGURE 3.6 Deep-water oxygen isotope record for the last 100 million years, illustrating the long-term cooling trend and the tendency for the system to evolve through abrupt transitions (arrows). (Adapted, by permission, from R.G. Douglas and F. Woodruff (1981). Copyright © 1981 by Cesare Emiliani.)

tions and whether there is a satisfactory explanation for abrupt changes. The committee recommends that research under this initiative fall into two basic types of case studies: (1) environments of extreme warm periods and (2) climate-biosphere connections during abrupt changes.

Environments of Extreme Warm Periods

The most important intervals to focus on in studying past warm periods in order to test the ability of models to simulate warmer climatic conditions are the Pliocene (3 to 5 Ma), the Early Eocene (50 to 55 Ma), and the mid-Cretaceous (100 Ma). For each of these intervals, tropical biota expanded into higher latitudes than at present and polar ice cover was greatly reduced. There may have been significant changes in oceanic circulation as well. Attempts have been made to model these warm climates (e.g., Barron, 1985; Barron et al., 1981; Crowley et al., 1986). The major paleogeographic changes of the last 100 million years are not sufficient to model the warm climates. Other factors such as increased atmospheric carbon dioxide may be needed to explain the warm periods.

Although each of the above intervals is important, the committee gives the Pliocene highest priority because it is the most recent time period for which we have evidence for climates significantly warmer than the present. In general, precision and resolution of climate data will be higher and sampling densities will be greater for more recent warm intervals than for older warm intervals. Pliocene flora and fauna are also very similar to modern, Pliocene records are widespread and easily accessible in both continental and marine settings, and most Pliocene records have undergone little alteration. All these features facilitate more quantitative estimates of environmental information and development of regional and even global patterns of climatic data. In addition, Pliocene warm intervals are punctuated by the abrupt development of wide-scale glaciation in the northern hemisphere at about 2.5 Ma (see the section "Climate-Biosphere Connections During Abrupt Changes" below).

Observational Needs. Observations are required

• to develop a global stratigraphy for warm intervals, particularly the Pliocene, that enables correlation of different sections on both land and sea. In order to implement this plan, substantial investments may be required to improve chronologies and acquire long records from terrestrial and marine environments.

• to derive quantitative estimates of a variety of climatic parameters and boundary conditions, including estimates of temperature and precipitation as well as seasonality of continental interiors. High-resolution time series and synoptic studies are required. Whenever possible, sampling intervals of time series should be fine enough to detect forcing functions on orbital time scales. At a minimum, synoptic studies should provide regional resolution comparable to model output. Additionally, proxy evidence for carbon dioxide levels and better assessment of paleorecords of wind, ice extent, and sea level are required. Changing boundary conditions such as oceanic gateways and orography must be better specified. Although individual research projects will be an important component in understanding past climates, fully successful implementation of this research plan and maximum interaction with models will require coordinated, multidisciplinary efforts that integrate work from a wide variety of disciplines and environments.

Modeling Needs. Efforts are required

• to experiment with atmospheric and oceanic models under radically different boundary conditions. These efforts will prove a sturdy test for climate models. For example, an oceanic GCM sensitivity experiment for an open Central American isthmus indicates that North Atlantic deep-water production may have collapsed (Figure 3.7). Since this same model will

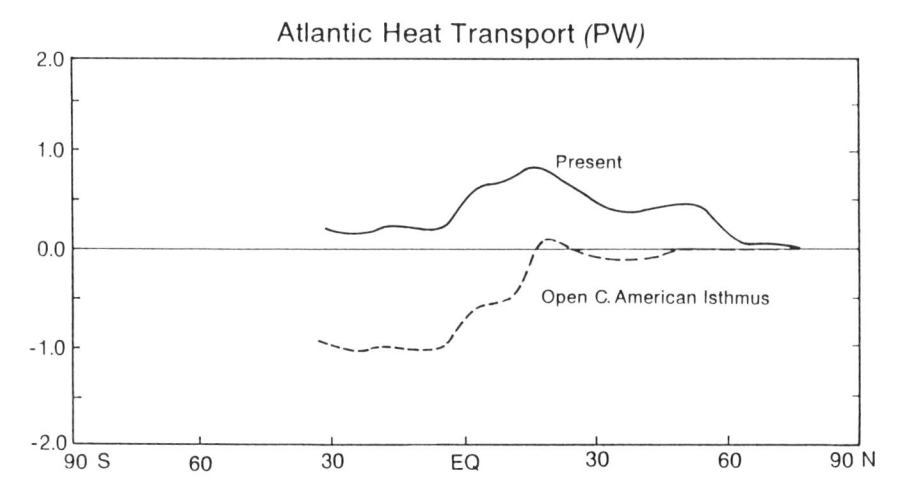

FIGURE 3.7 Ocean general circulation model experiment testing the effect of an open Central American isthmus on North Atlantic thermohaline circulation and poleward ocean heat transport (the latter is positively correlated with the amount of North Atlantic deep water produced). (Reprinted, by permission, from T.J. Crowley and G.R. North (1990). Copyright © 1990 by Oxford University Press: after Maier-Reimer et al., 1990.)

eventually be used to make greenhouse predictions, it is of interest to determine whether there is any geological support for such a large change. In fact, the record does provide some support for this conclusion (Woodruff and Savin, 1989; Delaney, 1990). Continued experiments with oceanic models under radically altered boundary conditions (Barron and Peterson, 1990) are needed. Other examples of modeling studies involve atmospheric GCMs and coupled oceanic-atmospheric models (e.g., Washington and Meehl, 1989). In addition, there is a need for further testing of models of warm, saline bottom-water production (Brass et al., 1982; Peterson, 1979). The geologic record represents the only realistic test for the parameterizations in these models.

Climate-Biosphere Connections During Abrupt Changes

A number of significant and relatively abrupt transitions have occurred during the last 100 million years (Figure 3.6). Some transitions, such as those near the Eocene-Oligocene boundary (34 Ma), appear to be closely associated with ice buildup and reorganization of oceanic circulation due to the tectonic evolution of ocean basins (Corliss and Keigwin, 1986; Kennett et al., 1974).

In general, the origin of abrupt transitions is not well understood. There are at least four classes of models exhibiting unstable behavior due to (1) thermohaline instabilities reorganizing the oceanic-atmospheric circulation (Broecker et al., 1985), (2) ice albedo feedback resulting in abrupt changes in ice volume (North and Crowley, 1985), (3) nonlinear feedbacks in the climate system leading to "internal" oscillations in climate (e.g., Saltzman and Sutera, 1984), and (4) carbon dioxide changes due to abrupt changes in ocean productivity (e.g., Arthur et al., 1988).

The importance of abrupt transitions for global change research is clear, as much of the concern about the human impact on the earth system involves the unprecedented rate of changes. Study of rapid transitions or abrupt changes between different states offers the potential to monitor effects of rapid change on the environment, including the response of the biosphere. Better understanding of these events will delineate the system response to a large, sudden perturbation.

A number of events stand out in their potential to contribute substantially to objectives of global change research. The two most promising intervals for study are the late Pliocene 2.5-Ma onset of mid-latitude northern hemisphere glaciation and the Eocene-Oligocene cooling (30 to 40 Ma) marked by expansion of antarctic ice and the largest biotic turnover in the Cenozoic. The Cretaceous-Tertiary (K-T) boundary event also merits attention in that it is a major extinction event associated with and likely caused by an asteroid or comet impact (Alvarez et al., 1980).

The development of extensive northern hemisphere ice sheets at about 2.5 Ma is a rapid transition from relatively ice free conditions in polar regions of the northern hemisphere. The 2.5-Ma event follows closing of the Isthmus of Panama, opening of the Bering Strait, and continued mountain building in Tibet and western North America (Ruddiman and Raymo, 1988). The cooling was accompanied by increasing aridity in the tropics and had a profound effect on life in the vicinity of the North Atlantic (Stanley, 1986). The emergence of the Isthmus of Panama at this time or slightly earlier separated marine organisms in the Atlantic and Pacific and allowed extensive interchange of mammals between North and South America.

The interval spanning the Eocene-Oligocene boundary marks the transition from the relatively warm periods of the early Cenozoic to the cold periods of the middle and late Cenozoic. The transition is marked by at least three relatively abrupt steps—the middle to late Eocene (40 Ma), the Eocene-Oligocene boundary (34 Ma), and the mid-Oligocene (30 Ma). These transitions are marked by some significant climate steps (Figure 3.6). There is evidence for ice sheet development as early as 40 Ma on Antarctica, and oxygen isotope evidence clearly records an additional large change at 34 Ma. There is additional evidence for glacial expansion on Antarctica about

30 Ma. The largest biotic turnover in the Cenozoic occurred around the same time as the Eocene-Oligocene transition. There were significant changes in marine organisms, terrestrial flora, and terrestrial vertebrates. The timing of some of these changes is still uncertain, along with their degree of abruptness and the cause of the overall pattern.

The end of the Cretaceous marks one of the most spectacular events in the earth's history—the probable impact of a 10-km bolide that is associated at least in part with widespread extinctions used to define the end of a geologic era. The discovery of the famous iridium layers in K-T sections has provoked some of the most stimulating geoscience research of the last decade. Although much has been learned about the K-T, a number of outstanding problems remain.

Observational Needs. Observations are needed

• to develop a comprehensive reconstruction of the physical changes in the environment across the abrupt events. Detailed time series are needed for key variables and in key regions in order to delineate the timing of the system response and the relationships between different components. The comprehensive reconstructions must focus on the physical, chemical, and biological state of the system at "snapshots" that span the abrupt event. The essential physical climate requirements are the distribution of surface temperatures, hydrologic state, seasonality of temperature and precipitation, cryosphere state, distribution and intensity of winds, and water mass distribution. Carbon dioxide levels are a key element of the chemical state of the system and the record of the carbon system, including productivity and carbon burial, are essential requirements.

• to develop a complete description of the distribution and character of the biosphere, including correlation of terrestrial floras, vertebrates, the faunas of the marginal marine environment and shelf, and planktonic and benthic organisms from the open ocean, in order to describe the ecological dynamics.

Modeling Needs. Modeling work is required

• to determine the origin of the abrupt transitions in the Cenozoic. The climate transition appears to reflect some type of instability in the climate system. However, it is not known what types of instability may have been involved and how they may have been triggered by long-term changes in continental position, orographic forcing, carbon dioxide, or ocean circulation. Modeling studies should focus on quantitative estimates of changes in these boundary conditions. A different type of modeling study should examine the effect of the long-term changes on more idealized models that can be used to explore properties of unstable systems. Additional modeling studies for the K-T should include examining the climate and chemical perturbations associated with asteroid impacts and large volcanic eruptions.

- to understand the relationship between abrupt environmental forcing and observed biotic responses. Although we have an approximate idea of the coincidence of the environmental and biotic transitions, the physical explanations for the biospheric response are still lacking.

CRITICAL PROGRAM ELEMENTS

Substantial progress in earth system history research has resulted from a proper balance between individual research projects and larger, coordinated efforts. In some instances the magnitude of the effort and the diversity of the expertise required to accomplish the objectives of the USGCRP will require research programs that are interdisciplinary, multiinstitutional, and international. However, the importance of maintaining smaller, single-investigator programs is also recognized.

To maximize further progress, some techniques and facilities need to be expanded and refined, and some new methods need to be developed. Examples include development and maintenance of facilities (e.g., drills) to acquire samples, calibration of the environmental records, and development of techniques and strategies to correlate diverse paleoenvironmental histories. Listed below are some of the major issues that must be addressed to implement the earth system history and modeling initiative in the USGCRP.

Sample Acquisition

The drilling of ice, lake, and ocean sediment cores will be the backbone of the proposed initiatives. Therefore drilling capability must be developed and maintained to meet anticipated needs. Producing high-temporal-resolution records is contingent on obtaining a sufficient amount of material to measure small sample volumes for multiple parameters. It is recommended that drills for collecting both marine sediments and ice cores be designed to take larger-diameter and higher-quality cores.

The committee has identified the following needs:

- ice cores (see the sections "The Last 1,000 to 2,000 Years," "Earlier Holocene Millennial-Scale Fluctuations," "Abrupt Change," and "The Last Glacial Cycle"). Ice cores are a treasure trove of information about past climates. The status of U.S. ice core drilling was reviewed recently, and recommendations were made (NRC, 1986) to develop and maintain a suite of drills for diverse programs in diverse areas. These may range from high (>18,000 feet), remote ice caps in the tropics and mid-latitudes to the polar ice caps, which are up to 3,000 m thick. Drilling these deep cores is expensive; however, it may be necessary to retrieve other long ice core records after the GISP II/GRIP drilling in Greenland (see the section "Abrupt Changes"). The committee recommends that this possibility be explored along with the potential of international cooperation (see the section "International Cooperation" below).

• long terrestrial records. Although we are making substantial progress in unraveling the history of the ocean basins, the terrestrial record is less well developed. Filling this gap requires expanding our capability to take long cores on land from sediments such as the thick loess sequences in China. Again, international cooperation on this project should be explored (see the section "International Cooperation").

• enhanced drilling of marine records (see the sections "Holocene High-Resolution Environmental Reconstructions," "Glacial-Interglacial Cycles," and "System Responses to Large Changes in Forcing"). The need for more marine cores will require additional drilling equipment and more efficient use of existing facilities. Currently, sediment sequences from the sea floor are recovered principally by the Ocean Drilling Program (ODP) or individual efforts by investigators on ships from oceanographic institutions. Meeting the needs of the USGCRP for long, continuous, high-resolution sequences will require some augmentation of drilling efforts. An enhanced ocean drilling effort dedicated to paleoceanography should allow for larger-diameter cores, multiple cores at each site, and perhaps the development of new capabilities for drilling very high sedimentation rate records in continental margins or sediment drifts. Cooperation with the ODP on this issue should be explored.

Environmental Calibration

The value of proxy records stems from our ability to reconstruct some aspect of the climate system. Implicit in this statement is that we understand the climatic signal embedded in the proxy. This requires careful study of modern conditions, as all proxies must be calibrated in terms of current conditions. Unfortunately, available data often are insufficient to perform this critical task. It is essential to allocate resources to the study of modern processes as an integral part of paleoclimatic and paleoenvironmental reconstructions and for refining and developing sets of modern observations to enhance proxy records.

The following is needed:

• process studies of modern environments. Quantitative specification of the physical and chemical processes creating the preserved proxy record must precede the development of empirical transfer functions used to extract paleoenvironmental information. Especially critical is explicit documentation of lags, thresholds, nonlinearities, and interactions between variables governing the responses preserved in paleoclimatic records (Graumlich and Brubaker, 1986). Examples of critical areas that would benefit from such process studies include fractionation of isotopes in precipitation, plankton, and tree rings; entrapment of gases within ice; and incorporation of trace metals into corals. Fairly long-term observations may be required in order to provide a statistically meaningful data set for calibration purposes.

• development of new proxy techniques to estimate environmental change. One major goal of the earth system history and modeling initiative is to provide quantitative estimates of environmental variabilities at key intervals of relevance to global change research. Achieving this requires expanding our capability to estimate such variables as temperature, salinity, and phosphorous. New geochemical methods may prove invaluable. For example, a new technique for estimating sea surface temperature or bottom-water temperature would be invaluable for separating the ice volume, salinity, and temperature signals from the oxygen isotope record in marine carbonates.

Correlation of Records

Before any definitive statements can be made about the climate at a certain time, samples must be temporally correlated with a high degree of accuracy. Current efforts to integrate different chronologies must be enhanced.

These efforts include

• enhanced capability for radiometric dating. A substantial number of accelerator carbon-14 dates will be required to accomplish the USGCRP goals. In addition, new applications of the cosmogenic isotopes (chlorine-36, beryllium-10) should provide valuable insights. Currently, AMS facilities exist at six institutions, and an additional facility is scheduled for completion in 1991 at Woods Hole Oceanographic Institution. Based on existing and planned facilities, Elmore et al. (1988) have identified a minimum annual shortage of approximately 4,000 non-carbon-14 AMS analyses to meet current program needs. The USGCRP will increase current demand for both carbon-14 and cosmogenic isotope measurements. The adequacy of existing U.S. facilities must be reassessed.

• extension and improvement of the current radiocarbon chronology. Separate efforts must be launched to update older, possibly erroneous measurements, as well as to extend the known radiocarbon chronology beyond that available from tree rings (9,600 B.P.).

• improvements in chronostratigraphic techniques. Beyond the range of carbon-14, additional techniques may be used to correlate paleoclimatic records. These include isotope stratigraphy, biostratigraphy, tephrochronology, and paleomagnetic stratigraphy. For example, land and sea records can be linked using pollen, dust, paleomagnetics, and tephrochronology. Although incremental advances in these areas are expected, a more focused research effort to improve these correlations may be required in some cases.

Data Management

Successful completion of many of the tasks outlined requires establishment of data bases ranging from those limited in scope to the needs of an individual project to global data sets needed for large international programs. The committee recommends that large-scale data bases be developed (1) when it becomes apparent that lack of organization is a deterrent to continued progress and (2) for projects requiring considerable coordination (e.g., the global network for the last 1,000 years (see the section "Global Network of Environmental Change"), the Little Ice Age (see the section "Little Ice Age"), and scenarios for greater warmth (see the section "Environments of Extreme Warm Periods")). The design of the data banks must be carefully considered by the project participants.

INTERNATIONAL COOPERATION

Informal international cooperation has led to abundant and fruitful scientific advances in studies of earth system history. Several of the proposed initiatives may require a more formal level of cooperation:

• acquisition of long ice cores. A number of groups in countries including the United Kingdom, Denmark, Switzerland, France, the USSR, and Australia might be interested in collaborating in efforts to drill and analyze these cores.

• acquisition of long land records with enhanced coring capability. Special priority should be given to collaborative research with the USSR and China, whose land masses account for such a large fraction of those in the northern hemisphere.

• acquisition of more long paleoceanographic records and development of new methods to take high-sedimentation-rate deep-sea cores. Progress in this area may require either close coordination with the ODP or establishment of separate arrangements. If coordination with the international ODP develops, it is essential to recognize that pre-Pleistocene studies currently are not part of the plans for the International Geosphere-Biosphere Program and that a U.S. connection with an international ODP is insufficient to ensure that pre-Pleistocene studies will be included.

• time slice reconstructions of past climates. These constitute major efforts that will require international participation and support. Discussions between U.S. and Soviet scientists are already under way for collaborative studies of the early Pliocene warming. The committee recommends expanded activities in this area.

In addition, the following are critical components of the USGCRP that could begin immediately at the international level: (1) coordination of existing data bases and sample collections, (2) planning for the coordination of new data bases, and (3) preliminary discussions of the scientific potential and logistical support necessary to mount large regional programs.

REFERENCES

Alvarez, L.W., W. Alvarez, F. Asara, and H.V. Michel. 1980. Extraterrestrial cause for the Cretaceous-Tertiary extinction. Science 208:1095-1108.

Arthur, M.A., W.E. Dean, and L.M. Pratt. 1988. Geochemical and climatic effects of increased marine organic carbon burial at the Cenomanian/Turonian boundary. Nature 235:714-717.

Barnola, J.M., D. Raynaud, Y.S. Korotkevich, and C. Lorius. 1987. Vostok ice core provides 160,000-year record of atmospheric CO_2. Nature 329:408-414.

Barron, E.J. 1985. Explanations of the Tertiary global cooling trend. Palaeogeogr. Palaeoclimatol. Palaeoecol. 50:45-61.

Barron, E.J., and W.H. Peterson. 1989. Model simulation of the Cretaceous ocean circulation. Science 244:684-686.

Barron, E.J., and W.H. Peterson. 1990. Mid-Cretaceous ocean circulation: Results from model sensitivity studies. Palaeoceanography 5:319-337.

Barron, E.J., S.L. Thompson, and S.H. Schneider. 1981. An ice free Cretaceous? Results from climate model simulations. Science 212(4494):501-508.

Beer, J., et al. 1988. Information on past solar activity and geomagnetism from [10]Be in the Camp Century ice core. Nature 331:675-679.

Berger, A.L., and L.D. Labeyrie. 1987. Abrupt climatic change—an introduction. Pp. 3-22 in W.H. Berger and L.D. Labeyrie (eds.), Abrupt Climatic Change. D. Reidel, Dordrecht, The Netherlands.

Berger, A.L., J. Imbrie, J.D. Hays, G. Kukla, and B. Saltzman (eds.). 1984. Milankovitch and Climate. D. Reidel, Dordrecht, The Netherlands. 895 pp.

Boyle, E.A. 1988. Vertical oceanic nutrient fractionation and glacial/interglacial CO_2 cycles. Nature 331:55-56.

Boyle, E.A., and L. Keigwin. 1987. North Atlantic thermohaline circulation during the past 20,000 years linked to high-latitude surface temperature. Nature 330(6143):35-40.

Brass, G.W., E. Saltzman, J.L. Sloan II, J.R. Southam, W.W. Hay, W.T. Holser, and W.H. Peterson. 1982. Ocean circulation, plate tectonics, and climate. Pp. 83-89 in Climate in Earth History (Studies in Geophysics). National Academy Press, Washington, D.C.

Broecker, W.S. 1982. Ocean chemistry during glacial time. Geochim. Cosmochim. Acta 46:1689-1705.

Broecker, W.S., and G.H. Denton. 1989. The role of ocean-atmosphere reorganizations in glacial cycles. Geochim. Cosmochim. Acta 53:2465-2501.

Broecker, W.S., D.M. Peteet, and D. Rind. 1985. Does the ocean-atmosphere system have more than one stable mode of operation? Nature 315:21-26.

Cane, M.A., S.C. Dolan, and S.E. Zebiak. 1986. Experimental forecasts of the 1982/83 El Niño. Nature 321:827-832.

Charlson, R.J., J.E. Lovelock, M.O. Andreae, and S.G. Warren. 1987. Oceanic phytoplankton, atmospheric sulphur, cloud albedo and climate. Nature 326:655-661.

CLIMAP Project Members. 1976. The surface of the ice-age earth. Science 191:1131-1144.

CLIMAP Project Members. 1981. Seasonal reconstruction of the earth's surface at the last glacial maximum. Map and Chart Series 36. Geological Society of America, Boulder, Colo.

CLIMAP Project Members. 1984. The last interglacial ocean. Quat. Res. 21:123-124.

COHMAP Members. 1988. Climatic changes of the last 18,000 years: Observations and model simulations. Science 241:1043-1052.

Corliss, B.H., and L.D. Keigwin, Jr. 1986. Eocene-Oligocene paleoceanography. Pp. 101-118 in K.J. Hsü (ed.), Mesozoic and Cenozoic Oceans. AGU Geodynamics Series 15:101-118.

Crowley, T.J. 1990. Are there any satisfactory geological analogs for a future greenhouse warming? Journal of Climate, in press.

Crowley, T.J., and G.R. North. 1990. Paleoclimatology. Oxford University Press, New York, in press.

Crowley, T.J., D.A. Short, J.G. Mengel, and G.R. North. 1986. Role of seasonality in the evolution of climate over the last 100 million years. Science 231:579-584.

Dansgaard, W., J.W.C. White, and S.J. Johnsen. 1989. The abrupt termination of the Younger Dryas climate event. Nature 339:532-534.

Delaney, M.L. 1990. Miocene benthic foraminiferal Ed/Ca records: South Atlantic and western equatorial Pacific. Palaeoceanography, in press.

Denton, G.H., and W. Karlén. 1973. Holocene climatic variations—their pattern and possible cause. Quat. Res. 3:155-205.

Dodge, R.E., R.G. Fairbanks, L.K. Benninger, and F. Maurrasse. 1983. Pleistocene sea levels from raised coral reefs of Haiti. Science 219:1423-1425.

Douglas, R.G., and F. Woodruff. 1981. Deep sea benthic foraminifera. Pp. 1233-1327 in C. Emiliani (ed.), The Sea, 7. Wiley-Interscience, New York.

Elmore, D., et al. 1988. Geoscience Research with New and Improved AMS Instrumentation. A Report of the Accelerator Mass Spectrometry Advisory Committee (AMSAC). 10 pp.

Fairbanks, R.G. 1989. A 17,000-year glacio-eustatic sea level record: influence of glacial melting rates on the Younger Dryas event and deep-ocean circulation. Nature 342:637-642.

Frenzel, B., and W. Bludau. 1987. On the duration of the interglacial to glacial transition at the end of the Eemian Interglacial (Deep Sea Stage 5 e): Botanical and sedimentological evidence. Pp. 151-162 in W.H. Berger and L.D. Labeyrie (eds.), Abrupt Climatic Change. D. Reidel, Dordrecht, The Netherlands.

Gaffin, S.R., M.I. Hoffert, and T. Volk. 1986. Nonlinear coupling between surface temperature and ocean upwelling as an agent in historical climate variations. J. Geophys. Res. 91:3944-3950.

Graumlich, L.J., and L.B. Brubaker. 1986. Reconstruction of annual temperature (1590-1979) for Longmire, Washington, derived from tree rings. Quat. Res. 25:223-234.

Guiot, J., A. Pons, J.L. de Beaulieu, and M. Reille. 1989. A 140,000-year continental climate reconstruction from two European pollen records. Nature 338:309-313.

Hansen, J., and S. Lebedeff. 1987. Global trends of measured surface air temperatures. J. Geophys. Res. 92(D11):13345-13372.

Harvey, D.L.D. 1988. Climatic impact of ice-age aerosols. Nature 334:333-335.

Hays, J.D., J. Imbrie, and N.J. Shackleton. 1976. Variations in the earth's orbit: Pacemaker of the ice ages. Science 194:1121-1132.

Hovan, S.A., D.K. Rea, N.G. Pisais, and N.J. Shackleton. 1989. A direct link between China loess and marine $\Delta^{18}O$ records: Aeolian flux to the north Pacific. Nature 349:296-298.

Huntley, B., and I.C. Prentice. 1988. July temperatures in Europe from pollen data, 6000 years before present. Science 241:687-690.

Huntley, B., and T. Webb. 1989. Migration: Species' response to climatic variations caused by changes in the earth's orbit. J. Biogeogr. 16:5-19.

Imbrie, J., et al. 1984. The orbital theory of Pleistocene climate: Support from a revised chronology of the marine $\Delta^{18}O$ record. Pp. 269-305 in A. Berger, J. Imbrie, J. Hays, G. Kukla, and B. Saltzman (eds.), Milankovitch and Climate. D. Reidel, Dordrecht, The Netherlands.

Imbrie, J., A. McIntyre, and A. Mix. 1989. Oceanic response to orbital forcing in the late Quaternary: Observational and experimental strategies. Pp. 121-164 in A. Berger, S.H. Schneider, and J.-C. Duplessy (eds.), Climate and Geosciences. Kluwer, Dordrecht, The Netherlands.

International Geosphere-Biosphere Programme (IGBP). 1989. Global Changes of the Past. Report No. 6. IGBP, Stockholm, Sweden. 39 pp.

Jacobson, G.L., Jr., T. Webb III, and E.C. Grimm. 1987. Patterns and rates of vegetation change during the deglaciation of eastern North America. Pp. 277-288 in W.F. Ruddiman and H.E. Wright, Jr. (eds.), North America and Adjacent Oceans During the Last Deglaciation. DNAG Vol. K-3. Geological Society of America, Boulder, Colo.

Jouzel, J., et al. 1987. Vostok ice core: A continuous isotope temperature record over the last climatic cycle (160,000 years). Nature 329:403-418.

Keeling, C.C., et al. 1989. A Three Dimensional Model of Atmospheric CO_2 Transport Based on Observed Winds: Observational Data and Preliminary Analysis. Appendix A in Aspects of Climate Variability in the Pacific and the Western Americas. Geophysical Monograph 55. American Geophysical Union, Washington, D.C.

Keigwin, L.D., Jr., and E.A. Boyle. 1985. Carbon isotopes in deep-sea benthic foraminifera: Precession and changes in low-latitude biomass. Pp. 319-328 in E.T. Sundquist and W.S. Broecker (eds.), The Carbon Cycle and Atmospheric CO_2: Natural Variations Archean to Present. Geophysical Monograph 32. American Geophysical Union, Washington, D.C.

Keigwin, L.D., Jr., and G.A. Jones. 1989. Glacial-Holocene stratigraphy, chronology, and paleoceanographic observations on some North Atlantic sediment drifts. Deep Sea Research 36:845-867.

Kennett, J.P., et al. 1974. Development of the Circum-Antarctic current. Science 186:144-147.

Knox, F., and M. McElroy. 1984. Changes in atmospheric CO_2: Influence of biota at high latitudes. J. Geophys. Res. 89:4629-4637.

Koerner, R.M. 1989. Ice core evidence for extensive melting of the Greenland ice sheet in the last interglacial. Science 244:964-968.

Kukla, G. (ed.). 1989. Long continental records of climate. Palaeogeogr. Palaeoclimatol. Palaeoecol. 72:1-225.

Kutzbach, J.E., and F.A. Street-Perrott. 1985. Milankovitch forcing of fluctuations in the level of tropical lakes from 18-O kyr BP. Nature 317:130-134.

LaMarche, V.C. 1974. Paleoclimatic inferences from long tree-ring records. Science 183:1043-1048.

LaMarche, V.C., and K.K. Hirschboeck. 1984. Frost rings in trees as records of major volcanic eruptions. Nature 307:121-126.

Legrand, M.R., R.J. Delmas, and R.J. Charlson. 1988. Climate forcing implications from Vostok ice-core sulphate data. Nature 334:418-420.

Luther, M.E., J.J. O'Brien, and W.L. Prell. 1990. Variability in upwelling fields in the northwestern Indian Ocean, 1, Model experiments for the past 18,000 years. Palaeoceanography 5:433-445.

Maier-Reimer, E., and U. Mikolajewicz. 1989. Experiments with an ocean GCM on the cause of the Younger Dryas. Report No. 39. Max-Planck-Institut fur Meteorologie, Hamburg, FRG.

Maier-Reimer, E., U. Mikolajewicz, and T. Crowley. 1990. Ocean GCM sensitivity experiment with an open Central American isthmus. Palaeoceanography 5:349-366.

Manabe, S., and R.J. Stouffer. 1988. Two stable equilibria of a coupled ocean-atmosphere model. Journal of Climate 1:841-866.

Martinson, D.G., N.G. Pisias, J.D. Hays, J. Imbrie, T.C. Moore, and N.J. Shackleton. 1987. Age dating and orbital theory of the ice ages: Development of a high-resolution 0-300,000-year chronostratigraphy. Quat. Res. 27:1-29.

Mayewski, P.A., G.H. Denton, and T.J. Hughes. 1981. Late Wisconsin ice sheets in North America. Pp. 67-178 in G.H. Denton and T.J. Hughes (eds.), The Last Great Ice Sheets. Wiley-Interscience, New York.

Mercer, J.H. 1978. West Antarctic ice sheet and CO_2 greenhouse effect: A threat of disaster. Nature 271:321-325.

Mesolella, K.J., R.K. Matthews, W.S. Broecker, and D.L. Thurber. 1969. The astronomical theory of climatic change: Barbados data. J. Geol. 77:250-274.

Mitchell, J.F.B., N.S. Grahame, and K.H. Needham. 1988. Climate simulation for 9000 years before present: Seasonal variations and the effect of the Laurentide Ice Sheet. J. Geophys. Res. 93:8282-8303.

Mosley-Thompson, E., L.G. Thompson, P.M. Grootes and N. Gundestrup. 1990. Little Ice Age (Neoglacial) paleoenvironmental conditions at Siple Station, Antarctica. Ann. Glaciol. 14:199-204.

National Research Council (NRC). 1986. Recommendations for a U.S. Ice Coring Program. National Academy Press, Washington, D.C. 67 pp.

Neftel, A., H. Oeschger, T. Staffelbach, and B. Stauffer. 1988. CO_2 record in the Byrd ice core 50,000-5,000 years BP. Nature 331:609-611.

North, G.R., and T.J. Crowley. 1985. Application of a seasonal climate model to Cenozoic glaciation. J. Geol. Soc. London 142:475-482.

Oeschger, H. 1985. The contribution of ice core studies to the understanding of environmental processes. Pp. 9-17 in C.C. Langway, Jr., H. Oeschger, and W. Dansgaard (eds.), Greenland Ice Core: Geophysics, Geochemistry, and the Environment. Geophysical Monograph 33. American Geophysical Union, Washington, D.C.

Oeschger, H., B. Stauffer, R. Finkel, and C.C. Langway, Jr. 1985. Variations of the CO_2 concentration of occluded air and of anions and dust in polar ice cores. Pp. 132-142 in E.T. Sundquist and W.S. Broecker (eds.), The Carbon Cycle and Atmospheric CO_2: Natural Variations Archean to Present. Geophysical Monograph 32. American Geophysical Union, Washington, D.C.

Peterson, W.H. 1979. A steady thermohaline convection model. Technical Report TR-79-4. Rosenstiel School of Marine and Atmospheric Science, University of Miami, Coral Gables, Fla. 160 pp.

Petit, J.R., L. Mounier, J. Jouzel, Y.S. Korotkevich, V.I. Kotylakov, and C. Lorius. 1990. Palaeoclimatological and chronological implications of the Vostok core dust record. Nature 343:56-58.

Pisias, N.G., J.P. Dauphin, and C. Sancetta. 1973. Spectral analysis of late Pleistocene-Holocene sediments. Quat. Res. 3:3-9.

Pokras, E.M., and A.C. Mix. 1987. Earth's precession cycle and Quaternary climatic changes in tropical Africa. Nature 326:486-487.

Prell, W.L. 1984. Monsoonal climate of the Arabian Sea during the late Quaternary: A response to changing solar radiation. Pp. 349-366 in A. Berger, J. Imbrie, J. Hays, G. Kukla, and B. Saltzman (eds.), Milankovitch and Climate. D. Reidel, Dordrecht, The Netherlands.

Prell, W.L., and J.E. Kutzbach. 1987. Monsoon variability over the past 150,000 years. J. Geophys. Res. 92:8411-8425.

Rind, D., and D. Peteet. 1985. Terrestrial conditions at the last glacial maximum and CLIMAP sea-surface temperature estimates: Are they consistent? Quat. Res. 24:1-22.

Rind, D., D. Peteet, W. Broecker, A. McIntyre, and W. Ruddiman. 1986. The impact of cold North Atlantic sea surface temperatures on climate: Implications for the Younger Dryas cooling (11-10k). Climate Dynamics 1:3-33.

Rind, D., D. Peteet, and G. Kukla. 1989. Can Milankovitch orbital variations initiate the growth of ice sheets in a general circulation model? J. Geophys. Res. 94:12851-12871.

Röthlisberger, F. 1986. 10,000 Jahre Gletschergeschichte der Erde. Aarau, Verlag, Sauerländer.

Ruddiman, W.F., and A. McIntyre. 1981. The North Atlantic Ocean during the last glaciation. Palaeogeogr. Palaeoclimatol. Palaeoecol. 35:145-214.

Ruddiman, W.F., and M.E. Raymo. 1988. Northern Hemisphere climate regimes during the past 3 Ma: Possible tectonic connections. Philos. Trans. R. Soc. London B318:411-430.

Saltzman, B., and A. Sutera. 1984. A model of the internal feedback system involved in late Quaternary climatic variations. J. Atmos. Sci. 41:736-745.

Sarmiento, J.L., and J.R. Toggweiler. 1984. A new model for the role of the oceans in determining atmospheric pCO_2. Nature 308:621-624.

Siegenthaler, U., and T. Wenk. 1984. Rapid atmospheric CO_2 variations and ocean circulation. Nature 308:624-625.

Stanley, S.M. 1986. Anatomy of a regional mass extinction: Plio-Pleistocene decimation of the western Atlantic bivalve fauna. Palaios 1:17-36.

Street-Perrott, F.A., and S.P. Harrison. 1984. Temporal variations in lake levels since 30,000 yr BP—an index of the global hydrological cycle. Pp. 118-129 in J.E. Hansen and T. Takahashi (eds.), Climate Processes and Climate Sensitivity. Geophysical Monograph 29. American Geophysical Union, Washington, D.C.

Stuiver, M., and T.F. Braziunas. 1988. The solar component of the atmospheric ^{14}C record. Pp. 245-266 in F.R. Stephenson and A.W. Wolfendale (eds.), Secular Solar and Geomagnetic Variations in the Last 10,000 Years. Kluwer, Dordrecht, The Netherlands.

Thompson, L.G., and E. Mosley-Thompson. 1987. Evidence of abrupt climatic change during the last 1,500 years recorded in ice cores from the tropical Quelccaya ice cap, Peru. Pp. 99-110 in W.H. Berger and L.D. Labeyrie (eds.), Abrupt Climatic Change. D. Reidel, Dordrecht, The Netherlands.

Thompson, L.G., E. Mosley-Thompson, W. Dansgaard, and P.M. Grootes. 1986. The Little Ice Age as recorded in the stratigraphy of the tropical Quelccaya ice cap. Science 234:361-364.

Thompson, L.G., et al. 1989. Holocene-Late Pleistocene climatic ice core records from Qinghai-Tibetan Plateau. Science 246:474-477.

Washington, W.M., and G.A. Meehl. 1989. Climate sensitivity due to increased CO_2: Experiments with a coupled atmosphere and ocean general circulation model. Climate Dynamics 4:1-38.

Webb, T., III, P.J. Bartlein, and J.E. Kutzbach. 1987. Climatic change in eastern North America during the past 18,000 years; Comparisons of pollen data with model results. Pp. 447-462 in W.F. Ruddiman and H.E. Wright, Jr. (eds.), North America and Adjacent Oceans During the Last Deglaciation. DNAG Vol. K-3. Geological Society of America, Boulder, Colo.

Webster, P.N., and N. Streten. 1978. Late Quaternary ice age climates of tropical Australia, interpretation and reconstruction. Quat. Res. 10:279-309.

Woodruff, F., and S.M. Savin. 1989. Miocene deep water oceanography. Palaeoceanography 4:87-140.

Zhang, J., and T.J. Crowley. 1989. Historical climate records in China and reconstruction of past climates. Journal of Climate 2:833-849.

4
Human Sources of Global Change

OVERVIEW

In its 1988 report the Committee on Global Change concluded that human interactions with the earth system should be a crucial component of the USGCRP. It recommended that the initial priority for research on such interactions should be to obtain a better understanding of the human sources of global change. Other interactions—in particular the human consequences and management of global change—were recognized as being of fundamental importance. But the most immediate requirement for progress in the overall, interdisciplinary global change program was felt to be a more systematic understanding of how human activities altered chemical flows, energy fluxes, and physical properties central to the operation of the earth system (NRC, 1988).

This chapter formulates a research plan for achieving a better understanding of the human sources of global change by the end of the decade

---

This chapter was prepared for the Committee on Global Change by Robert W. Kates, Brown University, Chairman; William C. Clark, Harvard University; Vicki Norberg-Bohm, Harvard University; and B.L. Turner II, Clark University, based on the input of the working group on Human Interaction with Global Change and a larger workshop that defined the highest-priority research areas and identified specific research needs and opportunities that could be accomplished within the next 5 years (see Appendix A for workshop participants). Members of the working group were Robert W. Kates, Brown University, Chairman; William C. Clark, Harvard University; Thomas Lee, Massachusetts Institute of Technology; V.W. Ruttan, University of Minnesota; Chauncey Starr, Electric Power Research Institute; B.L. Turner II, Clark University; and Vicki Norberg-Bohm, Harvard University.

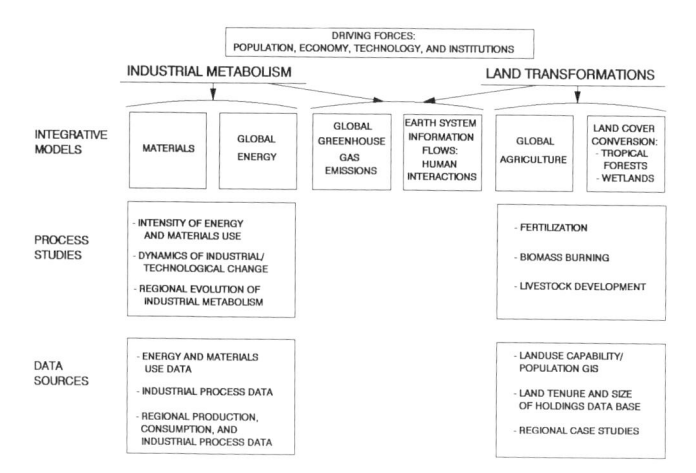

FIGURE 4.1 Recommended 1990-1995 research initiatives on human sources of global change.

and recommends priority research initiatives for implementation over the period from 1990 through 1995. In its deliberations on this issue the committee collaborated with a wide range of scholars and institutions (see Appendixes A and C), but worked particularly closely with the National Academy of Engineering's Technology and the Environment Program and the Social Science Research Council's Committee for Research on Global Environmental Change.

As shown in Figure 4.1, the recommended program focuses on two principal human sources of global change: industrial metabolism and land transformation. For each source, the program recommends research on integrative models, process studies, and data base development. In addition, a small number of synthesis studies are proposed.

Background

In its 1988 report the Committee on Global Change recommended the following research initiative on human interactions with the global environment (NRC, 1988):

> This research initiative would focus on the relatively short term record of the period of intensive human activities that have affected the global environment. Anthropogenic changes in the earth system need to be systematically documented over the past several hundred years and analyzed as a basis for developing useful reference scenarios of future change. In particular, two aspects of human activity are especially relevant to global change: land use changes, which influence both physical (e.g., albedo, evapotranspiration, and

trace gas flux) and biological (e.g., vegetative cover and biodiversity) variables; and the industrial metabolism that transforms resources into emission that must be absorbed and processed by the environment.

The committee also made a recommendation related to building the scientific foundations for research into the human dimensions of global environmental change. It recognized that research in the human dimensions of global change was relatively underdeveloped and thus there was a need to support discipline-oriented research in this area, with "the relevant research communities encouraged to develop their own internally justified research priorities relevant to global change." This is discussed further in the section "Investigator-Initiated Research" (below).

After completion of the committee's 1988 report, "the green book," the Working Group on Human Interactions with Global Change was formed, with the specific task of further defining a research agenda in the areas of land transformations and industrial metabolism. This group developed an initial list of priority research topics and planned a larger workshop to bring together active researchers and representatives of related institutional efforts (Appendix C). In preparation for this workshop, a background literature review was prepared, as well as several brief statements describing key directions for future research (Appendix B). The workshop was attended by 28 people from a wide range of disciplinary backgrounds (Appendix A). This group was asked to define the highest-priority research areas and to identify specific research needs and opportunities that could be accomplished within the next 5 years. This chapter draws heavily on the discussions at that workshop and on written material submitted by workshop participants.

Priority Recommendations

From among the research initiatives discussed in this chapter, the committee selects eight to receive the highest and immediate priority. To develop analytical frameworks through integrative models of human sources of emissions and land cover change, the committee recommends foci on

- global agriculture,
- global greenhouse gas emissions, and
- regional land cover conversion, particularly in tropical forests and wetlands.

To support the model development with key studies of important processes, the committee recommends foci on

- fertilization in agriculture,
- biomass burning, and
- intensity of energy and materials use.

To provide critical data for model development and process studies, the committee recommends

- a common information system (geographic information system) for land use, land cover, or land capability and population density and
- a data base for global historical energy and materials use.

THE RESEARCH PROGRAM

Over the next 5 years the human interactions research program focuses on the agricultural, energy, and industrial activities that generate the crucial materials flow and land use changes that are responsible for most human-induced global change. The intent is to describe the distribution of these activities in detail appropriate for incorporation into the evolving earth system models and to understand the forces in society that drive them well enough to create alternative projections of these activities over the time frame of a century. Beyond this immediate task the research effort should provide the foundation for subsequent studies of the impacts of human-induced change and efforts to control or to adapt to such change.

Three general levels of scholarship are required in this effort: data collection, process studies, and synthesis.[1]

Data collection: Data collection has both a historical and a current component. It includes data on human activities that lead to changes in the chemical flows, physical properties, and surface covers of interest, as well as data on demographic, technical, and socioeconomic variables.

Process studies: Two types of studies fall under the heading "process." They are most easily distinguished by the following simple model of global environmental change.

| changes in chemical flows, physical properties, or surface cover | = | changes in emissions or conversions per unit human activity | × | changes in mix and level of activities |

The first type of process study describes the emissions and conversions per unit human activity. Currently, emission coefficients for the transformation of materials and energy are better understood than those for the transformation of land. Many of the coefficients for industrial processes, such as carbon dioxide emissions for various energy technologies, are well documented. In contrast, the coefficients for land use processes, such as methane emissions from various rice cultivation techniques and animal husbandry, are not well understood.

The second type of process study describes the mix and level of activities (supply and demand) that transform the environment. Examples of this type of process study include studies on the determinants of the level of electricity consumption or agricultural production; studies that shed light on the interrelationships between demand for various energy technologies, or for different crops or farming practices; and models of future levels of energy or agriculture production.

Synthesis: Finally, in the area of synthesis, our ultimate goal is to improve the capacity to describe a range of potential futures for the environmental components of interest. Thus we are looking here for models or analytical frameworks that allow for the development of consistent future scenarios of human activities that force global environmental change. Synthesis is required to combine the two types of process studies, and to develop approaches that integrate the changes from land use and industrial metabolism.

INDUSTRIAL METABOLISM

The term "industrial metabolism" has come in recent years to signify the total pattern of energy and materials flows through an industrial sector or region. The basic goal of research in this area is to understand how changing levels of human population, economic activity, technology, and social organization influence the pattern of such flows—in particular the output of pollutant chemicals relevant to global change.

Integration and Synthesis

The committee identifies two integrative modeling goals in the area of industrial metabolism. The first, relatively well in hand, is the creation of global models of energy use and associated pollutant emissions. Such models have a long history of development in studies of the greenhouse effect. They have provided internally consistent and plausible scenarios at the decade-to-century scale of future carbon dioxide emissions under a range of policy and development assumptions.[2]

In recognition of the importance of activities other than energy use as key contributors to climatic change, the need now is to develop models that provide information on other gases of interest and thus include, in addition to energy use, industrial activity and agriculture. One such effort that is under way is a greenhouse gas model based on a general equilibrium framework. Because this model synthesizes across both land use and industrial activity, it is discussed in the section "Earth Systems Information Flow Diagram for Human Interactions" (below).

The second goal is to develop models that focus on the transformation of materials through industrial processes. These models would be concerned not only with greenhouse gas emissions, but also with other industrial emissions that contribute to global environmental change. This is a more complex task at a much earlier stage of development. One approach to this type of modeling is under way at the Institute for Economic Analysis at New York University. This effort builds on previous work on a world input-output model. This analysis will incorporate detailed technical process information and provide quantities and geographic distribution of pollutant emissions under various scenarios as one of its outputs. "The objective of the proposed study is to identify and evaluate concrete, consistent, economically feasible strategies for environmentally sound development, that is, to examine alternative approaches to reducing poverty over the next 50 years while also reducing global pollution" (Duchin, 1989).

Process Studies and New Data

Three areas of process studies were singled out as being specifically needed to support the modeling work noted above. First, in certain regions and industrial sectors the amounts of energy and material being used per unit value of production are rising, while in others they are falling. Needed is a better understanding of the human factors responsible for these patterns. Second, research has shown that the emergence of lead technologies such as textiles, iron and steel, chemicals, and electronics has in the past led to radical transformations in the overall character of industrial metabolism. These historical cycles of technology dominance suggest the emergence of one, perhaps two, new lead technologies over the next century, with major implications for emissions and waste streams. The factors that determine the timing of these long-term transformations need to be understood. Third, changes in industrial metabolism seem to occur as integrated regional phenomena, rather than on a sector-by-sector basis. Needed is a deeper understanding of the economic, technological, and institutional processes that determine such regional integration.

There is a need for data base developments related to each of the process studies noted above. Global historical data are needed on the changing intensities of energy and material use across a range of human activities. Also required are global long-term records of the changing pattern of emissions from major transforming technologies. Finally, it will be necessary to document integrated histories of industrial input-output characteristics for selected regions around the world.

The Intensity of Energy and Materials Use

Materials intensity and energy intensity are defined as the quantity of material or energy consumed per unit of value created (e.g., per unit GNP or per unit end-use service) or as the quantity of material or energy consumed per capita. Studies seeking to understand the trends in energy and materials intensities are critical to an understanding of the human forcing of global environmental change because these intensities are one of the key determinants of the environmental insult caused by economic or industrial activity.

There are three previous studies of particular relevance to this research area. The first is a study of trends in materials use, which concludes that the United States is indeed experiencing a trend toward dematerialization, i.e., a reduction in the quantity of material or energy consumed per unit of value (Williams et al., 1987). This work is based on about 100 years of data on the consumption and prices of the following materials: steel, cement, paper, ammonia, chlorine, aluminum, and ethylene, as well as several low- and intermediate-volume metals. This study concludes that dematerialization is the result of a structural shift in the United States based on the level of income. It also concludes that dematerialization may cause the rate of growth in U.S. industrial demand for energy to become zero, or even negative. A second study based on the U.S. data documents trends in U.S. energy intensity over the past 100 years. It concludes that the availability of abundant and low-cost electricity played an important role in the decline of energy intensity (measured as energy per unit GNP) that the United States has experienced since 1920 (Schurr, 1984).

Both of these studies provide important starting points. However, there is a need to examine other countries, both in similar and in different stages of development, to further identify and understand the trends in energy and materials use. Furthermore, as emphasized by a third study, it is necessary to consider what measures of dematerialization are meaningful with regard to the environment (Herman et al., 1989). This study suggests that dematerialization be defined as the amount of waste generated per unit industrial product and that distinctions need to be drawn between the dematerialization of production and that of consumption. It also suggests that several noneconomic factors are relevant to trends in dematerialization.

To gain a better understanding of the determinants of energy and materials intensity, new data need to be developed, analyzed for trend, and placed within a broader framework of understanding economic development. Long-term historical surveys of energy and materials use, by country, are needed. For energy, this would include data disaggregated by end use, fuel mix, energy carrier (electricity, steam loop, or on-site combustion), and the combustion or generating technology. For materials, the task is to determine the relevant materials to include in the data base and the metric(s) to be used in accounting for them. Candidate materials of particular environmental concern include metals, paper, plastics, chemical commodities, and fertilizers. Accounting schemes could be based on mass, as well as some other measure of environmental insult. In the case of materials, information on both the industries or economic sectors that produce the materials and the intermediate and final consumer is needed. In addition, it is important to properly account for the import and export of materials and energy, both in their raw state, and as embodied in products.

With a historical record of material and energy requirements for a number of countries at different stages of economic development available, analysis to understand the causes and implications of observed trends should follow. An important focus will be on the changes in energy or materials requirements for providing end-use services.[3] This is of particular interest, as the well-being of societies is based on end-use values rather than primary inputs. Thus if a good or service can be supplied with a lower energy or materials intensity, the environmental insult can be lowered without any reduction in well-being. In addition, end-use analysis is critical for understanding what is technically possible and therefore for developing future scenarios of materials and energy use.

In examining the causes of observed trends in materials and energy intensity, key factors to consider include stage of economic development, level of GNP, rate of growth of GNP, level and rate of change of economic productivity, product life cycle, and the demography of infrastructure. The relevant economic data required for this analysis, such as prices, capital and labor inputs, and productivity are generally available. After an understanding of the causal factors underlying historical trends in energy and materials intensity is developed, the next step will be to explore the implications this has for the future by constructing plausible scenarios of future levels of energy and material use.

A final task would be exploratory in nature. On the basis of the insights gained from the above analysis, it would try to develop a theory or conceptual framework to explain the relationship between trends in materials and energy use and economic development. Specifically, this theoretical work will describe the relationship between energy and materials intensities and stages of economic development, the relationship between changes in energy and materials intensity and changes in economic productivity, and the amount of energy or materials required to support economic growth.

The Dynamics of Industrial and Technological Change

"The two themes of ecological and economic interdependence are strongly linked through a third theme—the development and diffusion of technology" (Brooks, 1986). Because of this linkage, improved understanding of the factors that influence the development and diffusion of technology will significantly improve our understanding of the human forcing of global environmental change and improve our ability to develop future scenarios of emissions from industry. Of particular relevance is research aimed at understanding what factors influence the timing of technological change at three levels of analysis: industry, sociotechnical systems, and technological eras.

At the industry level, the effort will focus on case studies of specific industries that have particularly large environmental impacts, either through high energy use (e.g., aluminum manufacturing) or through emissions from specific manufacturing processes (e.g., smelters). The industries with the largest environmental effects include electric power generation, chemical manufacturing, mining, mineral processing (including metal working), and paper manufacturing.

The case studies of specific industries will include detailed information on the technologies in use, including all resources used in the production process and the waste streams created. Particular attention will be paid to technological changes that affect the environment, including materials substitution, the efficiency of materials and energy use, source (waste) reduction technologies, and recycling. The data will also include economic indicators of industry and individual firm performance. Data will be historical, documenting the changes in each industry's technological and economic characteristics over the past 150 years. Data also need to be collected on promising new technologies for each industry examined.[4]

At the level of sociotechnical systems, previous empirical work suggests that there are long-term regularities in the evolution, diffusion, and replacement of these systems (i.e., there is a characteristic time constant for substitution between technologies).[5] The goal of this effort is to understand the reasons for this time constant, why it varies across countries, and whether it varies over time. Studies are needed both at the level of specific industries and for major sociotechnical systems such as energy or transportation systems. Factors to consider in this analysis include economic growth, factor abundance and productivity, and absolute wealth, as well as institutional and cultural factors.

Of particular interest are the ways in which individuals, firms, and governments influence the timing and direction of technological change. The study would seek to identify the most influential factors in the decision-making process at each of these levels, the relative importance of each set of factors in influencing technological change, and the importance of the interactions between these groups.

Finally, previous research at the level of the technological eras has shown that lead technologies such as textiles, iron and steel, chemicals, and electronics characterize major eras, as do their power sources of water, steam, electricity, and gasoline. Conflicting and somewhat controversial theories exist that purport to explain the assembly and differences of these lead technologies and their associated economic impacts but little has been done to examine their environmental import (Ayres, 1989). Of particular interest for global change is that the historical cycles of technology dominance suggest the emergence of one, perhaps two, new lead technologies over the next century, with major implications for emissions and waste streams. This project would identify candidates for new lead technologies and their implications for emission coefficients and energy and systems intensity.

Regional Evolution of Industrial Metabolism

A proven approach for analyzing industrial metabolism is the materials balance method. This method is based on the concept of conservation of mass. It tracks the use of materials and energy "from cradle to grave." In other words, it follows them from extraction through manufacturing, consumption, and disposal, and then to their final environmental destination. It is a tool that allows economic data to be used in conjunction with technical information on industrial processes, the use and disposal of products by consumers, and environmental transport to describe chemical flows to the environment.[6] This type of analysis was used to study the Hudson-Raritan river basin. This study reconstructed the emissions of heavy metals and other chemical wastes for the past 100 years (Ayres et al., 1988). Two important conclusions were drawn from this analysis: (1) Major sources of environmental pollutants have been shifting from production to consumption processes. (2) Large numbers of materials uses are inherently dissipative, spreading widely into the environment.

For comparison purposes, the next study should be of a river basin in a developing country. Possible candidates include the Ganges in India and the Zambezi in Zimbabwe and Mozambique. This would allow for a preliminary comparison of the environmental impacts of industrialization for different stages of development and under differing development paths. It would also contribute further to an understanding of how the sources of environmental perturbations shift between production and consumption during the development process. This project would require the development of two types of data bases. The first is a data base of economic statistics on production and consumption. This data base must be sufficiently disaggregated, both geographically (regional rather than country level statistics) and by end use. The second is a data base of the relevant industrial processes for the region of study.

LAND TRANSFORMATIONS

Throughout most of history, the primary way in which humans have effected change in the global environment has been by transforming the earth's land surface. In many less industrialized regions, it remains so today. Land transformation, in the sense used here, includes not only activities like forest clearing that change land cover type and physical properties, but also those like fertilizer use that change chemical flows.

Integration and Synthesis

The committee identifies two integrative modeling initiatives in the field of land transformation. The first is the creation of a global model of agricultural activities and their associated impacts on land use, land cover, earth surface properties, and chemical flows. Such a model would play a role in global change studies directly analogous to that already served by existing energy models that forecast carbon dioxide emissions. In particular, it would relate changes in human populations, economic activity, technology, and institutional structures to changes in the extent of agricultural lands under various cropping regimes, in irrigation practices, in fertilizer use, and in other activities directly relevant to global change studies. The model should be global in scope, with a century forecasting horizon and regional resolution. A second group of models should address transformations in two land uses that the committee has determined to be of particular relevance to global change studies: tropical forests and coastal wetlands. In both cases, the goal would be to provide regional forecasts of the impact of human activities on the extent of the affected land uses over decade-to-century scales.

Global Agriculture Model

Agriculture (including pastoralism) is the human activity most closely associated with land transformation as well as a major source of nitrogen, phosphorus, and methane. Existing models that project global agricultural activities rarely extend beyond 25 to 40 years and have been designed to answer questions of demand and trade in agricultural products or food security rather than as major sources of important emissions or spurs to land conversion.[7]

A new generation of models is required that can develop internally consistent scenarios of the expansion of agricultural production over the next century to meet the needs of a world population of 10 billion. The major goal of this new modeling effort will be to estimate the bounds of environmental change emanating from agricultural land use changes, including earth surface properties and chemical flows. Such a model would thus track the inputs required for the intensification of agriculture, including nitrogen and phosphorous from fertilizers, irrigation, and pesticides and herbicides as well as land transformations between agriculture and other uses.

In order to accomplish this, the model must incorporate resource opportunities and constraints (agroecological considerations) and socioeconomic trajectories (e.g., demand, technology, and institutions) for the major facets of agricultural change (e.g., land expansion and intensification). In other words, the model must adequately represent the relationship between technical and institutional factors, land transformations, the global environmental impact of those transformations, and the impact on the local, national, and international economies. The model must provide regional resolution representing technical, economic, and institutional factors faced by different countries, particularly the current differences in agricultural development between the developed and the developing nations. Other requirements of the model include (1) that it be sufficiently open that all parameters are subject to examination and change by the user and (2) that it be sufficiently generic to enable the user to respond to changes in the problems that will need to be addressed as perceptions of local and global agricultural production and environmental assessment change over time.

Land Cover Projections for the Twenty-first Century

Two types of land cover have been identified as especially important because of the degree of change that they are likely to experience and the impacts that these changes will have on global environmental systems. These are forests and wetlands.

Forest conversion includes deforestation and afforestation and focuses on tropical forests. This focus is taken because of the high rates of deforestation throughout the tropics and the projected negative environmental impacts of this phenomenon—global warming, species loss, sedimentation, soil degradation—and because of the complexity of forces that are giving rise to it—population pressures, technological capacity to deforest or afforest and transport resources, and international demands for tropical products.

Wetland conversion involves the loss and gain of lands with standing water owing to such phenomena as drainage, wetland agriculture (e.g., field raising), irrigation, ponding, and reservoirs. Wetland conversions, particularly rice cultivation, are important in global change because of their impacts on atmospheric chemistry (such as methane), biodiversity, and water availability and will be driven by the need to feed and provide water for the additional 5 billion people projected on earth by the year 2050, most of whom will be situated in the tropical realms where wet rice cultivation is climatically suitable.

Understanding of the rates and trajectories of these changes will follow from matching studies on the proximate sources of change (see the section "Process Studies" below) with studies on the two land cover changes. The latter will be developed by detailing the changes for a number of selected regions or countries that account for a large share of the conversion in question. For example, studies of forest conversion will focus on the transformations in Amazonia, Borneo and the Malay Peninsula, and one or two other cases. Study of wetland conversion will focus on selected cases in South and Southeast Asia, West Africa, and perhaps Central America.

With data on the rates, trajectories, and processes of change, a number of scenarios for each of the cases of forest and wetland conversion will be developed. These scenarios will involve differing assumptions about the rates and trajectories of change as they are affected by changes in the driving and mitigating forces of change (e.g., international markets, population, conservation laws, and national park protection) and by estimates of the recoverability of the environment in question. One scenario for each case will focus on surprise changes, unsuspected rates or trajectories of change, and impacts of changes on the rates.

These scenarios aim (1) to develop the bounds and trajectories for the future and the consequent impacts that they will have on the global environmental systems through chemical emissions and through physical and biological changes and (2) to demonstrate the relative importance of the driving and mitigating forces of change by varying situations (and also time-space scales), including the synergisms among them, such that the impacts of policy can be projected.

Process Studies

The committee identifies three land transformation processes for which deeper understanding is urgently required to promote the overall goals of global change research. The first is fertilization. Needed is a better understanding of the human determinants of rates and character of fertilizers in major agroecosystems of the world. A second process requiring immediate attention is biomass burning. How do population density, land tenure patterns, economic development opportunities, and other human factors affect the frequency, character, and extent of biomass burning in major agroecosystems? Finally, better understanding is needed of livestock development. At the decadal scale, what large social forces control changes in the extent, character, and location of livestock utilization by humans?

Fertilization

Fertilization refers to the use of synthetic fertilizers to increase agricultural output, with possible expansion to include herbicides and pesticides. While this includes inputs to sustain grasslands or pastures, emphasis will be placed on the use of nitrogenous fertilizers associated with the intensification of production, particularly with the use of high yielding varieties. Currently, more nitrogen is fixed synthetically than naturally, with fertilizer use growing at 3 to 4 percent per year globally.[8] The increased use of synthetic fertilizers has direct impacts on emissions to the atmosphere, water quality, albedo, and biomass and indirect impacts on alternative land uses and methane emissions through the use of wet rice production. The increase in fertilization is worldwide and occurs in very different sociotechnical contexts. The driving forces for fertilization in different societies require much more detailed understanding if they are to be projected over the long term. This study will examine fertilizer use trends in their differing sociotechnical contexts.

Biomass Burning

Fire, as reflected in biomass burning, is estimated to account for 25 to 30 percent of anthropogenic carbon dioxide emissions to the atmosphere.[9] In addition, it accounts for a significant portion of the emissions of other greenhouse gases, including methane, nitrous oxides, nitrogen oxide, and carbon monoxide (U.S. EPA, 1989). It is a critical element of landscape transformation because it is important to permanent and cyclical forest clearance, to the maintenance and regeneration of grasslands and some tree species, and as a noncommercial fuel supply. Regardless of its purpose, under all circumstances it changes the albedo of the burn zone and increases particulates released to the atmosphere, and in many cases it leads to a loss in biodiversity, soil quality, and water retention. The reasons for its use and the impacts that it has, of course, vary by circumstance. Broadly speaking, three types of uses can be recognized—land use expansion, land intensification, and land maintenance. These can result from market pressures, local food pressures, lack of alternative fuels, and so forth. This study will situate biomass burning within these contexts and link burning to standardized measures, such as number of fire hours per annum, perhaps segregated into low- and high-intensity burns.

Livestock Development

Since the mid-1970s, livestock populations have risen by 6 percent worldwide, but by 11 percent in Africa and 19 percent in Latin America, with a subsequent increase in methane production and deforestation (WRI and IIED, 1988). Thus a perhaps overlooked element of change has been the expansion and intensification of livestock production, particularly in the developing world. These include the frontier expansion of pasture in Latin America and the intensification of livestock production within traditional systems, near urban areas for commerce, and within urban areas to meet the needs of the urban poor. The impacts on the environment are both immediate and indirect. Frontier expansion typically involves deforestation and land conversion to crop and pasture, while intensification involves increased demands for food and fodder, which affects land use and quality. In either case, albedo and biodiversity changes follow. An overall increase in livestock also increases emissions from their waste. The reasons for livestock development go beyond commercial demand and local needs; in parts of Latin America, social status is gained from livestock ownership, and in Africa, cattle ownership is an integral part of social relationships. Studies of this source of global change must differentiate between cases of expansion and cases of intensification and identify the related environmental changes and forces that give rise to livestock development.

Data Needs

Three data base developments are required in support of a better understanding of land use transformation as a human source of global change. First is a capability to compare population census data, land cover data, land use data, and land capability data within a common information system. Second is the development of a global data base on land tenure and size of holdings—two variables of importance in determining patterns of land transformation. Finally, there is a need to assure that regional case study data collected in various aspects of the global change program dealing with land transformation are complementary to the maximum extent possible. The danger is an uncoordinated approach in which ecological data are collected at one site, tropospheric chemistry data at another, and human activity data at a third. Instead, plans should be advanced for the establishment of long-term regional resource sites where relevant studies from the natural and human sciences can be conducted in concert, and their data sets pooled.

Land Use, Land Cover, or Land Capability/Population Density Geographic Information System

The universal measure of potential human activity is the number of persons within a unit area. Repeated censuses, usually on a decadal basis, take place in most countries of the world, and their accuracy exceeds that of most other global data sets (errors have been estimated at less than 3 to 20 percent). Many flows and land use changes seem directly proportional to population changes, and for others population may be the best available short-term surrogate for potential human impact. But population data are normally collected by administrative or political divisions of national territory and thus do not relate directly to land use or capability. Thus the first candidate for a human interaction data set is an integrated land cover, land use, or land capability/population density geographic information system (GIS) at a resolution of 10,000 to 100,000 km^2. Such work should merge the different global land surface data sets with new population census results as they emerge and maintain and update the resulting GIS in an appropriate national or international institution. Once developed, this GIS should also integrate data from industrial censuses around the world that can identify the geographic location of industries by four-digit standard industrial classifications.

Land Tenure and Size of Holdings

In order to pursue studies into the causes of various land transformations, a data base of standardized information on land tenure and land holdings is needed. Currently, the best resource for these data is the Wisconsin Land Tenure Center at the University of Wisconsin in Madison. For over two decades, this center has been collecting land tenure data for parts of Latin America, the Caribbean, Africa, and Asia. Although these data are not available in tabulated or electronic forms, original data sources are kept in their library. This proposed project would include both the continued collection of land tenure and land holding data and the development of a readily available, standardized data base, one that would be related to the GIS for land cover and population.

Regional Case Studies and Research Centers

The IGBP working group on data and information systems has identified a set of case studies in which efforts will be made to integrate remotely imaged and field-observed land use data. At the same time, investigator-initiated research projects on integrated case studies of global change and human impacts and response, including the Critical Environmental Zones project in the United States, have been funded in several countries. The early regional case studies should be brought together to encourage parallel or joint natural and human science studies in the same regions. These should serve as pilot efforts that lead to the development of regional research sites where research on long-term patterns of social, economic, and ecological change can be brought together.

INTEGRATIVE STUDIES ACROSS LAND USE AND INDUSTRY

Our ultimate goal is not just to understand industrial metabolism or land transformation, but rather to understand the ways in which human activities in general force changes in the earth system. Although it is too early in the overall research program to tackle this ultimate goal directly, three broad synthesis studies that could usefully be undertaken in the period from 1990 through 1995 are recommended by the committee. These initiatives are a global model of greenhouse gas emissions, a diagram of earth system information flows for human interactions, and an analysis of the driving forces of human-induced global change.

Global Model of Greenhouse Gas Emissions

The first generation of global models applied to the greenhouse problem addressed a single human activity—energy use—and a single greenhouse gas—carbon dioxide. The need is now clear for a second generation of models that address all the major gases and all the major human sources arising from both industrial and land use activities. Initial work has begun on such models and should be encouraged as part of the USGCRP.

Desirable characteristics of the new generation of models include (1) disaggregation by country or region; (2) disaggregation by human activities leading to emissions, including energy, agriculture, manufacturing, transportation, and services; (3) provision of a link between the details of technology (engineering or microlevel studies) and the macroeconomy; (4) modeling of interactions between the different human activities leading to emissions of greenhouse gases (e.g., the availability of biomass for energy use is dependent on agricultural technology and institutions); (5) output in 5- or 10-year steps, for 100 years; and (6) inclusion of institutional arrangements as one determinant of technological choices.

A "second-generation model" of greenhouse gas emissions, which represents an improvement over the current Institute of Energy Analysis/Oak Ridge Associated Universities model is being developed at Battelle Pacific Northwest Laboratories (Edmonds and Reilly, 1983; Edmonds et al., 1988). The new model will be an improvement in the following respects: modeling of all greenhouse gases rather than only carbon dioxide; a general equilibrium rather than partial equilibrium analytical structure; greater disaggregation of human activities, including agriculture, energy, transport, manufacture, and services; interactions between managed and unmanaged ecosystems; and improved modeling of resources, turnover in capital stocks, and international trade. Another approach is the input-output analysis described in the section "Industrial Metabolism" (above). While this analysis is not aimed solely at greenhouse gas emissions, it will provide information on greenhouse gas emissions in addition to other pollutant emissions.

Earth Systems Information Flow Diagram for Human Interactions

The global change program has benefited tremendously from the early effort of the Earth Systems Sciences Committee (ESSC) to create a "wiring diagram" of key process connections and information flows among the climatic, biogeochemical, and ecological components of the earth system. The ESSC effort, however, treated humans as a external boundary condition, rather than as an integral component of the system. Scholarship on the human dimensions of global change is now sufficiently advanced that the diagram should be revised to reflect our conceptual understanding of the interactions between people and environment that constitute the earth system.[10]

Driving Forces: Population, Economy, Technology, and Institutions

Human sources of global change arise from the needs of populations and their economies and technologies and are mediated by their institutions. These driving forces of human sources are widely recognized by scientists of society and technology, who differ in emphasis and causal attribution given to each factor. Enough is already known about the historical and geographic variation in driving forces to expect that a set of quite varied clusters of driving forces will emerge related to major regional and historical differences in population, economy and technology.

For example, the Earth as Transformed by Human Action study of 13 key pollutant emissions and land use conversions over the past 300 years suggests three varied trajectories in which population, economy, and technology each exercise successively greater influence as driving forces of these sources (Turner et al., 1990). Regional case studies suggest that all three trajectories are at work in the world today. Thus many places can be found in the world today with carbon emissions per unit area that are similar but driven by quite different combinations of forces. Further systematic study of these relations is needed.

IMPLEMENTATION REQUIREMENTS

Related Institutional Efforts on Human Interactions with Global Change

As discussed above, the research agenda defined in this chapter is sharply focused on obtaining a better understanding of the human forcing of global environmental change. This work is complemented and supported by the work of several other organizations that are involved in developing research agendas or sponsoring research into human interactions with global environmental change. By far the largest domestic effort is being carried out in the executive branch agencies. Information on executive branch initiatives can be found in the CEES document (CES, 1989). The activities of other organizations are briefly described in Appendix B, and the names of persons to contact for further information are included.

Two organizations deserve particular mention as their agenda in part shares the focus of this committee. The Social Science Research Council has formed a Committee for Research on Global Environmental Change, which is examining six areas, one of which is land use. The National Academy of Engineering shares the interest of the Committee on Global Change in industrial metabolism as part of its Technology and the Environment Program.

Investigator-Initiated Research

A strong program of investigator-initiated research is necessary to complement the highly focused nature of the research program outlined in this chapter, concentrating as it does on the human sources as inputs to the evolving model of the earth system. It is necessary that investigators be able to define studies on the impacts of global change on human activities and on societal efforts to control or to adapt to global change.

As stated in *Toward an Understanding of Global Change* (NRC, 1988),

> The existing research program on the human components of global change is also inadequately developed, as discussed in the background paper on the human dimension. Efforts to bring together natural, social, behavioral, and engineering scientists to examine in depth the research required on the human dimension of global change should be supported. Several research areas identified in the background paper—integrated methods to assess the risk and implications of long-term environmental change for resource availability at the regional scale; ways that knowledge, perceptions, and values related to global change can be more effectively brought to bear on human choices that affect global change; and evaluation and design of institutional mechanisms for better management of global change—require further development in close collaboration with those relevant scientific communities in the social, behavioral, and engineering sciences that were not adequately represented in current planning activities.

Currently, a major source of investigator-initiated research on human interactions is the National Science Foundation interdisciplinary program in human dimensions of global change. In the initial round of awards in 1989, there were 35 applications and 10 awards for $0.75 million. Funding was awarded to 4 investigator-initiated research projects as well as 6 programs for conferences, workshops, and institutionally based committees that were developing research agendas or research programs on global environmental change. These projects were of high scientific merit, diverse, and imaginative, spanning such topics as law and the transformation of water rights, social learning in the management of global environmental risks, critical zones in global environmental change, equity issues and the global greenhouse effect, and the economic impacts of global environmental change. A rapid expansion of this program is envisaged over the next 5 years.

Education and Training

The human sciences, social and behavioral, have special requirements for education and training. Global change phenomena are not at the core of their disciplines (except for some anthropology and geography) as they are for the earth and ecological sciences, and the normal scientific program of incentives and support works poorly in mobilizing scientific effort on this vital concern. Thus special emphasis needs to be given to encouraging younger social and behavioral scientists to participate in global change research, to enable them to obtain training and research experience, even outside their own institutions and across disciplinary lines, if needed. A key feature for such an effort would be a predoctoral fellowship program, including cross-institutional internships, and a postdoctoral interdisciplinary program. Such a program should be institutionally based, so that it can both provide essential training and support the development of the emerging centers of human interaction research in universities, institutes, and national laboratories.

Data Preparation and Dissemination

It is not too early to begin to plan for data preparation and dissemination. For the data projects outlined in this study, it is important that data be systematically archived, prepared, and disseminated at a central site and made readily accessible to interested researchers. These activities might be integrated within an IGBP Regional Research Center, at one of the national laboratories, or in the emergent regional research centers described in the section "Regional Case Studies and Research Centers" (above). Oak Ridge National Laboratory has competence and experience in such archiving and dissemination through their handling of the Carbon Dioxide Information Analysis Center (CDIAC).

THE STEPS BEYOND

The initial research program on human interactions with global change will almost surely be overtaken by events: critical observations of ongoing global change and rapidly increasing interest in impacts and policy research. Over the 5-year period of this focused research effort, it will become necessary to expand into these evolving research areas. The committee acknowledges and anticipates such expansion by emphasizing fundamental as well as focused research, an infrastructure for human interactions research, and close ties with the other initiatives of the behavioral and social sciences research community to understand and to respond to the challenge of global change.

NOTES

1. These three levels of scholarship correspond to the scientific objectives used by FCCSET's Committee on Earth and Environmental Sciences (CEES; formerly the Committee on Earth Sciences) of documentation (observation and monitoring), improved understanding, and development of (predictive and conceptual) models. The FY 1990 initiatives for the human dimensions of global environmental change as specified by the CEES are listed below:

(1) Data Base Development. This consists of two projects: (a) Land Surface Data Systems. Provide for the permanent archiving, management, access, and distribution of land surface earth science data sets for global change research on the interaction between human activities and environmental processes. (b) Improvement of Social Data Systems. Improve data resources dealing with individual and institutional actions affecting environmental changes.
(2) Understanding Processes of Change. Fundamental research on the relationships among global and environmental changes and human activities, including social, economic, political, legal, and institutional processes.
(3) Modeling Processes of Human Interactions with the Environment. Initial methodological and substantive research to develop more sophisticated models of human and institutional interactions in global change.

2. These models are generally economically based models, or are of the engineering-economic type. In addition, there have been some efforts to integrate the macro- and micro-levels of analysis. A thorough review of these energy models in relationship to global environmental change can be found in Toth et al. (1989). These models are also reviewed in Appendix C of this report. The most widely used economically based model is the IEA/ORAU model (Edmonds and Reilly, 1983). A recent example of the engineering-economic type model is described in Goldemberg et al. (1988). The most recent modeling efforts that integrate the macro- and micro-levels are described in U.S. EPA (1989).

3. Of several studies of this type in the area of energy, the most notable recent study was done by Goldemberg et al. (1988).

4. There exists today a family of empirically based engineering-economic energy and mass-balance models of conventional and advanced coal-to-electric conversion technologies. Two major efforts toward the development of process data bases were undertaken in the late 1970s and early 1980s, one at Statistics Canada and one at the International Institute for Applied Systems Analysis (IIASA). These are reviewed in Gault et al. (1985). The current recommendation differs from these previous efforts in one important way; the interest here is in a process data base that not only describes current processes but also is both historical and forward-looking.

5. Two sources of this work are Marchetti (1983) and Marchetti and Nakicenovic (1979). For a review of this work, see Ausubel (1989).

6. For a discussion of this methodology, see Ayres and Rod (1986) and Ayres et al. (1989).
7. For reviews of existing models, see Toth et al. (1989).
8. Based on data for 1974 to 1976 and 1983 to 1985 (WRI and IIED, 1988).
9. How, W.M., M.H. Liu, and P.J. Crutzen. Estimates of Annual and Regional Releases of CO_2 and Other Trace Gases to the Atmosphere from Fires in the Tropics, Based on FAO Statistics, 1975 to 1980. Presented at Symposium on Fire Ecology at Freiberg University, Federal Republic of Germany, May 16-20, 1989. Proceedings to be published by Springer-Verlag, Berlin.
10. An initial contribution toward this goal can be found in *Climate Impact Assessment*, which explores the interaction of society and climate.

REFERENCES

Ausubel, J.H. 1989. Regularities in technological development: An environmental view. In J.H. Ausubel and H.E. Sladovich (eds.), Technology and Environment. National Academy Press, Washington D.C.

Ayres, R.U. 1989. Technological Transformations and Long Waves. Research Report 89-1. International Institute for Applied Systems Analysis, Laxenburg, Austria.

Ayres, R.U., and S.R. Rod. 1986. Reconstructing an environmental history: Patterns of pollution in the Hudson-Raritan Basin. Environment 28(4):14-20, 39-43.

Ayres, R.U., L.W. Ayres, J.A. Tarr, and R.C. Widgery. 1988. An Historical Reconstruction of Major Pollutant Levels in the Hudson-Raritan Basin: 1880-1980. NOAA Technical Memorandum NOS OMA 42. United States Department of Commerce, National Oceanic and Atmospheric Administration, Washington, D.C.

Ayres, R.U., V. Norberg-Bohm, J. Prince, W.M. Stigliani, and J. Yanowitz. 1989. Industrial Metabolism, the Environment, and Application of Materials-Balance Principles for Selected Chemicals. Research Report 89-11. International Institute for Applied Systems Analysis, Laxenburg, Austria.

Brooks, H. 1986. The typology of surprises in technology, institutions, and development. In W.C. Clark and R.E. Munn (eds.), Sustainable Development of the Biosphere. Cambridge University Press, New York, N.Y.

Committee on Earth Sciences (CES). 1989. Our Changing Planet: The FY 1990 Research Plan. Office of Science and Technology Policy, Federal Coordinating Council on Science, Engineering, and Technology, Washington, D.C.

Duchin, F. 1989. Project Proposal: Strategies for Environmentally Sound Development: An Input-Output Analysis. Institute for Economic Analysis, New York University, New York.

Edmonds, J.A., D.F. Barns, and W.U. Chandler. 1988. Modeling Future Greenhouse Gas Emissions. Pacific Northwest Laboratory, Washington, D.C. September.

Edmonds, J., and J. Reilly. 1983. Global energy and CO_2 to the year 2050. The Energy Journal 4(3).

Gault, F.D., R.B. Hoffman, and B.C. McInnis. 1985. The Path to Process Data. Futures 17:509-527.

Goldemberg, J., T.B. Johansson, A.K.N. Reddy, and R.H. Williams. 1988. Energy for a Sustainable World. Wiley Eastern Limited, New Delhi.

Herman, R., S.A. Ardekani, and J.H. Ausubel. 1989. Dematerialization. In J.H. Ausubel and H.E. Sladovich (eds.), Technology and Environment. National Academy Press, Washington, D.C.

Marchetti, C. 1983. The Automobile in a System Context: The Past 80 Years and the Next 20 Years. Research Report 83-18. International Institute for Applied Systems Analysis, Laxenburg, Austria. July.

Marchetti, C., and N. Nakicenovic. 1979. The Dynamics of Energy Systems and the Logistic Substitution Model. Research Report 79-13. International Institute for Applied Systems Analysis, Laxenburg, Austria.

National Aeronautics and Space Administration (NASA). 1988. Earth System Science: A Closer View. Report of the Earth Systems Sciences Committee. NASA, Washington, D.C.

National Research Council (NRC). 1988. Toward an Understanding of Global Change: Initial Priorities for U.S. Contributions to the IGBP. National Academy Press, Washington, D.C.

Schurr, S.H. 1984. Energy use, technological change, and productive efficiency: An economic-historical interpretation. In Annual Review of Energy 1984. Annual Reviews Press, Palo Alto, Calif.

Toth, F.L., E. Hizsnyik, and W.C. Clark (eds.). 1989. Scenarios of Socioeconomic Development for Studies of Global Environmental Change: A Critical Review. Research Report 89-4. International Institute for Applied Systems Analysis, Laxenburg, Austria.

Turner, B., R.W. Kates, and W.C. Clark. 1990. The great transformation. In B. Turner, R.W. Kates, and W.C. Clark (eds.), The Earth as Transformed by Human Action. Cambridge University Press, New York, in press.

U.S. Environmental Protection Agency (EPA). 1989. Policy Options for Stabilizing Global Climate. Draft Report to Congress. EPA, Office of Policy, Planning and Evaluation, Washington, D.C. February.

Williams, R.H., E.D. Larson, and M.H. Ross. 1987. Materials, affluence, and industrial energy use. In Annual Review of Energy 1987. Annual Reviews Press, Palo Alto, Calif.

World Resources Institute (WRI) and the International Institute for Environment and Development (IIED). 1988. World Resources 1988-1989. Basic Books, New York.

5

Water-Energy-Vegetation Interactions

OVERVIEW

Over the last decade, considerable progress has been made in understanding and modeling both the climate system in general and certain related components of the terrestrial biosphere (e.g., biophysical-atmospheric exchanges such as radiation and water and heat fluxes, ecosystem dynamics, and trace gas exchange and biogeochemical cycles). As yet, however, there have been few successful efforts, either in modeling or in data acquisition, to link the activities of the physical climate and the terrestrial biosphere so as to further understand and improve capabilities to predict global change.

This chapter focuses on the interactions between the vegetated land surface and the atmosphere, particularly on the exchanges of energy, water, sensible heat, and carbon dioxide between the two. The aim of the research strategy discussed here is not merely to describe such exchanges but to fully understand them so that predictive modeling can be used to explore possible future states of the earth system. To do this, it will be necessary to make comprehensive, biophysically based models of the atmosphere and land biosphere with measurable state parameters as prognostic variables, e.g., temperature and humidity for the atmosphere, and albedo, leaf area index, and

---

This chapter was prepared for the Committee on Global Change from the contributions of Piers J. Sellers, University of Maryland, Chair; John Bredehoft, U.S. Geological Survey, Christopher Field, Carnegie Institute of Washington; Inez Fung, NASA/Goddard Institute for Space Studies; Alan Hope, San Diego State University; Gordon McBean, University of British Columbia; and William Reiners, University of Wyoming.

photosynthetic capacity for the land. To calibrate, initialize, and validate these models, it will be necessary to acquire a broad range of data covering the space-time domain, from continuous global-scale monitoring to intense, high-resolution field observations. Briefly stated, the goals of the research strategy are as follows:

• To develop models that realistically describe the interaction between the land biota and the atmosphere with particular reference to the exchanges of energy, water, heat, and carbon dioxide. Ultimately, these models should be adequate for exploring the consequences of global change in terms of perturbations to the climate system and terrestrial ecosystems. The models will have to cover a wide range of spatial scales (millimeters to global) and time scales (seconds to millennia).

• To collect data that can be used to initialize, validate, and prescribe boundary conditions for the models described above. Additionally, the data are to be used for monitoring the global environment and testing new hypotheses and for diagnostic or retrospective studies.

• To conduct manipulative experiments, field campaigns, and process studies to improve our understanding of the processes controlling the transfer of energy, water, heat, and carbon dioxide between the land surface and the atmosphere at appropriate scales and to develop better methods for quantifying the controls.

Figure 5.1 shows the relationship between the modeling, the process studies (experiments and laboratory work), and the large-scale data acquisition program proposed here, and the interdependence and coordination needed among these activities. In general, the modeling studies are intended to distill the results of process studies to provide a realistic predictive capability. The model predictions can be compared with field experiment observations on local scales and with data from the long-term monitoring effort and global data sets on regional and global scales, respectively. The flow of information should be two-way, as the results from the modeling activities will determine which variables should be observed at which time and space scales.

To achieve these goals, two things must be done. First, the existing research activities must be adjusted and focused to promote the coordinated, interdisciplinary studies necessary for the collection of specialized data sets and the construction of a new generation of models. Second, efforts must be undertaken to address the most crucial resource as of now, which is not hardware or data, but trained scientists.

The remainder of the chapter addresses the particulars of data needs, modeling, infrastructure, and resources, with particular emphasis on those areas that are weakly supported at present (see Tables 5.1, 5.2, and 5.3 for activities in monitoring, manipulative experiments, and field experiments,

WATER-ENERGY-VEGETATION INTERACTIONS

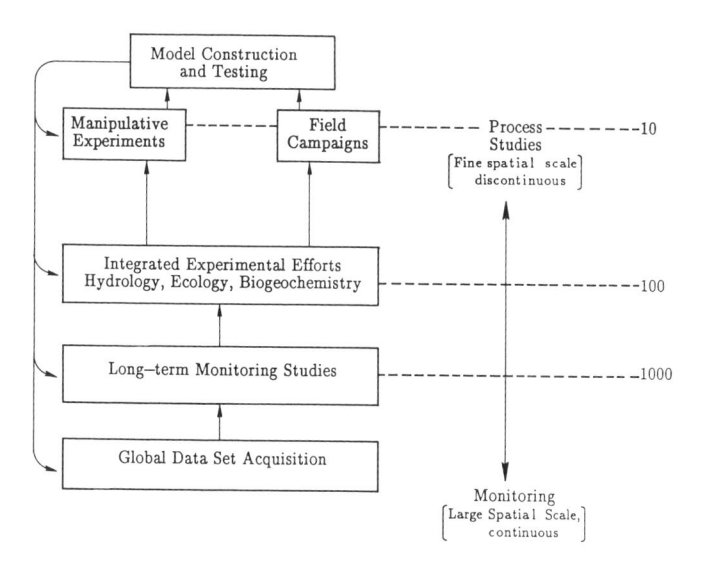

FIGURE 5.1 Relationship between modeling, process studies, and other data acquisition activities. The numbers on the right-hand side of the figure denote the approximate number of field sites worldwide.

respectively). The sections on data needs and modeling follow the structure outlined in Figure 5.1 by first reviewing global data set needs and then reviewing data needs at progressively finer spatial and temporal scales. It should be remembered throughout that the proposed strategy was conceived as a whole and that coordination of the component activities represents a significant challenge by itself.

DATA NEEDS AND EXPERIMENTS

A wide range of data needs to be gathered, processed, and integrated to provide an information base for modeling and diagnostic studies of the earth system. These include satellite, atmospheric, and in situ observations operating on an extensive and more-or-less continuous basis to provide "monitoring" information, and focused, coordinated observations—the product of field and in vitro experiments—to provide the insight necessary for model and algorithm development. These two kinds of data sets should be regarded as complementary, as the monitoring data sets will partially determine the list of items to be addressed by experiment and the experimental results should lead to improvement of the data processing methodologies applied to the monitoring data set.

It should be remembered that all types of land cover need to be ad-

TABLE 5.1 Monitoring Programs (Ongoing)

Activity	Program	Agency/Country
Meteorological data	Climate and global change program	NOAA, national meteorological agencies
Satellite data products	National Environmental Satellite, Data, and Information Service and weather service	NOAA, national meteorological agencies
Earth radiation budget	Earth Radiation Budget Experiment	NASA
Cloud climatology	International Satellite Cloud Climatology Project	NASA
Vegetation index (AVHRR)	GIMMS; National Environmental Satellite, Data, and Information Service; and other national programs	NASA, NOAA
Retrospective land cover studies	IRAP/International Satellite Land Surface Climatology Project	NASA
Soil moisture, vegetation cover	Goddard Space Flight Center	NASA
Snow and ice	Goddard Space Flight Center	NASA, NOAA
Runoff	?	WCRP
Carbon dioxide	Background Air Pollution Monitoring Network	NOAA

TABLE 5.2 Manipulation Experiments/Gradient Studies (Ongoing)

Activity	Program	Agency
Carbon dioxide enrichment, small crops	Florida Area Cumulus Experiment	DOE/USDA
Plastic ecosystem response, remote sensing	Experimental lakes area	Canada, universities
Effect of land use changes on hydrology (forests)	Experimental lakes area Valdai, RSFSR	Canada USSR
Effect of land use changes on hydrology (forests) and nutrient treatments	Hubbard Brook, Coweeta	USA
Crop and pasture fertilization	Worldwide	Agricultural agencies

TABLE 5.3a Field Campaigns (Ongoing)

Activity	Date	Program	Location	Lead Agency/ Country
Large-scale hydrometeorology $(100 \text{ km})^2$	1986	HAPEX	France	France
Energy balance, biophysics, meteorology, remote sensing $(15 \text{ km})^2$	1987, 1989 1988 on	FIFE KUREX	Kansas, USA Kursk, USSR	NASA USSR

TABLE 5.3b Field Campaigns (Planned)

Activity	Date	Program	Location	Lead Agency/ Country
Global Energy Water Cycle Experiment	1990s	Global Energy and Water Cycle Experiment	Global	WMO
Arid zone, energy, water cycle, remote sensing $(100 \text{ km})^2$	1992	HAPEX-II	Niger	France
Arid zone, energy, water cycle, remote sensing $(15 \text{ km})^2$	1991	IFEDA	Spain	European Community
Tropical forest, energy, water cycle, remote sensing	1990s	ABRACOS	Brazil	U.K., Brazil
Boreal forest, energy, water cycle, tropospheric chemistry, remote sensing	1994	BOREAS	Canada	NASA, Canada

dressed by the data acquisition effort and experiments, not just "natural, undisturbed" ecosystems. In particular, agricultural systems require a high priority as targets for studies at all levels—monitoring, process studies, and modeling. Integration with national agricultural research programs will be essential.

Particular needs for each kind of data set are discussed below.

Global Data Needs

Global data are needed for the specification of boundary conditions for global models as well as for the framework for analyzing and detecting global change. The data would come from extensive ground and aircraft surveys as well as from satellites. It should be stated at the outset that for satellite data to be useful for detecting and monitoring interannual and longer-term changes (1) calibration of satellite data must be of the highest priority, (2) rigorous correction for atmospheric effects and for viewing geometry must be performed, and (3) the calibration and correction procedures must be carried out by data centers and information systems to make the products available to the scientific user community in a timely fashion.

In terms of looking for global change indicators, it should be remembered that a large archive of satellite data (20 years' worth) already exists. More resources should be made available to study these records to determine whether changes or the effects of changes can be detected at a usable level of accuracy.

The following data are required:

• Satellite data. The most obvious need is for the integration of existing techniques into information systems that can deliver calibrated, geometrically corrected, atmospherically corrected, and registered data products to the scientific user community in a timely fashion. Until this is done, remotely sensed data will continue to be underused, if not unused, as research scientists are forced to complete the whole task themselves—from satellite sensor counts all the way through to derived products and only then toward their own scientific goals. Better means must be found.

It is proposed that more than one effort be initiated to address this task. To be effective, individuals involved will have to form a wide pool of talents and be drawn from a diverse population—instrument engineers, atmospheric physicists, scientific user groups, and information scientists—and will also have to work across agencies. They will have to develop a means for selecting and implementing satellite data algorithms for *operational* processing of the data stream. The development of these information systems is probably the highest-priority task facing the community at present.

The research on satellite data sets involves the continuing improvement of techniques for obtaining area-averaged parameters from coarse-resolution data. More research needs to be done to apply scaling methods so that knowledge gained from infrequent high-resolution data (e.g., Landsat) can be applied to the continuously acquired global coverage data sets (e.g., Advanced Very High Resolution Radar (AVHRR)).

- Land transformation data. Any kind of large-scale land transformation needs to be documented in a uniform way as part of the monitoring program. If possible, this should be connected with the satellite data acquisition effort.
- Topographic data. At present there are no reliable high-resolution topographic data sets for the globe. The resolution of available global data sets for the land surface is either 5 or 10 minutes, depending on the continent. Present data are inadequate for many land surface studies. There is a need (1) to compile, archive, and make available additional existing topographic data; (2) to prepare new topographic maps from space-borne (e.g., Système Probatoire d'Observation de la Terre (SPOT)) data; and (3) ultimately to fly a dedicated mission (or missions) to acquire a coherent set of topographic data that can be made available to earth scientists in readily usable forms.
- Vegetation data. There are several digital data sets of vegetation for the globe. The spatial resolution ranges from $0.5° \times 0.5°$ to much coarser for the globe. The resolution of vegetation information is much coarser, and ranges from about 10 biomes to more than 150 vegetation types. Some of these vegetation data sets represent potential or climax vegetation, and others include land use and modification. The accuracy of the data sets has to be improved. The feasibility of land surface classification using satellite data has now been demonstrated. Improvement of this new technique in conjunction with ground surveys is critical, not just for mapping vegetation, but also for developing the capability for detecting changes in the structure and function of vegetation.

It should be recognized that the appropriate grouping of vegetation types for evapotranspiration and water cycling may be different from the appropriate grouping for, say, surface energy balance. Research into appropriate classification schemes is needed.

- Vegetation function data. The phenology of vegetation is important for determining the timing and amount of water released through the land surface. There is limited tabular information on leaf area indices for different vegetation types. The normalized difference vegetation index (NDVI), available from polar orbiting satellites, gives information on the seasonal march of vegetation greenness globally and should be applied to the water studies. In order that the NDVI be used for detecting interannual and longer-term changes in phenology and water cycling, calibration of satellite instruments and rigorous correction for atmospheric effects must be of the highest priority. These studies should be carried out in conjunction with surface validation efforts (see following sections).

- Soil data. Most global soil data sets are digital versions of the Food and Agriculture Organization (FAO) soil maps. Improvements to soil data sets are on the research agenda of several international groups (e.g., Soil and Terrain Data Base (SOTER) of the International Soil Science Society). Soil units, texture, and other parameters in the FAO soil maps are important for hydrologic studies, as are hydraulic potential and rooting depth of vegetation. Some information about these parameters has been gleaned from the literature, but an extensive survey of different vegetation-soil complexes should be carried out.
- Soil moisture data. The temporal and spatial variations of soil moisture are not well known. In most models, the soil-water-holding capacity associated with different vegetation types is specified, but there is no observed climatology of the field capacity. Field capacity climatologies derived as residuals from budget calculations suffer from inaccuracies in the other terms in the water budget, such as rainfall and evapotranspiration. The capability for measuring soil moisture from space, especially in the presence of vegetation, must be developed as far as possible. Meanwhile, because it is impossible to map soil moisture distributions globally, surface-based studies should be carried out to understand the vegetation and soil characteristics that determine soil moisture amount and its spatial variations (see the section "Remote Sensing" below).
- Meteorological data. Near-surface atmospheric humidity, rainfall, near-surface air temperature, and wind speed are the driving forces for atmosphere-surface fluxes of water vapor. At present, the weather station network is the primary source for such data. Such networks must be maintained at mid-latitudes and expanded in polar and tropical regions. Technologies for automating these weather stations exist and should be further explored and developed, including such technologies as telemetric tipping rain gauges and ceilometers. Commonly, these data are used by the National Meteorological Center and then discarded. Efforts should be made to integrate some of these data into a research data base.
- Rainfall data. Rainfall varies on small scales. It also exhibits a distinct diurnal cycle, which varies geographically. Quantifying and understanding this variability are important to the scaling-up of local site studies to the gross resolution of global models. Thermal infrared radiation at the top of the atmosphere has been shown to be a useful index of convective precipitation, which is a source of small-scale variability. Existing satellite data (e.g., from AVHRR) should be analyzed to investigate the spatial and temporal variability of precipitation and to determine the appropriate precipitation statistics to describe rainfall variability on a global scale.

Microwave measurements of precipitation from satellites offer a real hope for providing a global rainfall climatology. More efforts should be committed to research and validation in this area.

- Runoff data. River runoff determines the freshwater inputs to the world oceans. Broecker (1989) has suggested that changes in runoff may trigger changes in deepwater formation and consequently changes in climate. The USGS has an extensive gauge network that measures, among other things, daily flow rates of the rivers and tributaries in the United States. Similar gauge data have yielded a long-term record of the flow of the Amazon River. Such networks must be maintained, and if possible, expanded to other major rivers of the world.

For closed drainage basins (see the section "Integrated Monitoring and Process Studies" below), river flow data must be analyzed in conjunction with contemporaneous precipitation and other weather data to develop and test hydrological models and to determine the frequency and accuracy of river flow data for global change.

- Surrogate and corroborative data. As there are several sources of water vapor to the atmosphere, the unique signatures of each water exchange process will provide information to validate models and hypotheses of the terrestrial hydrological cycle. These signatures include oxygen and hydrogen isotopes of water. Also, for C_3 plants, the simultaneous exchange of hydrogen oxide and carbon dioxide through stomata makes the temporal variation of carbon dioxide and its stable isotopes carbon-13 and oxygen-18 critical cross-checks for evapotranspiration.

- Photometric data. Standardization, coordination, and augmentation of sun photometry measurements are needed to allow the collection of atmospheric optical thickness data. These data are required for the routine atmospheric correction of satellite data.

Long-Term Monitoring

A global network of minimally instrumented sites is needed to provide data for

- diagnosing the effects of climatic change over long periods of time,
- testing basic models and hypotheses, and
- anchor stations to perform satellite algorithm inversion.

The first objective, diagnosing the effects of climatic change, will require different measurement strategies in different parts of the world. These effects may manifest themselves as changes in characteristics such as vegetation cover or composition, mass and heat fluxes, or the hydrological balance. Sites need to be identified that can act primarily as "barometers" that will reflect the impact of climatic shifts on different ecosystems while also providing data for the second objective, the long-term validation of models. This large, diffuse network of stations should total several hundred sites worldwide, and therefore optimal use will need to be made of existing resources and monitoring programs such as the Long-Term Ecological Research (LTER) network or the proposed observatories of IGBP and WCRP. The range of sites should encompass the world's major vegetation types and biomes, including agricultural systems.

The long-term monitoring stations should be located at the centers and edges of ecosystems. This arrangement would provide data that may indicate the resilience of ecosystems to changes as well as the shifts on the borders where initial changes could be expected. The full benefit of the long-term data sets will be realized only if sufficient resources are set aside for basic research that will analyze the data sets. Furthermore, it will be essential to provide adequate funds for technical personnel to maintain good quality data collection programs over an extended period.

Long-term monitoring of selected hydrological, climatological, and chemical variables has been conducted by a variety of agencies in the past. A reexamination of these monitoring efforts is required with a view to extending and coordinating the monitoring activities. Some of the automated data collection devices may be modified to record additional variables.

At minimum, these sites should be committed to collecting meteorological data and conducting periodic surveys of the vegetation and soils in their locale.

Integrated Monitoring and Process Studies

Out of the larger set of long-term monitoring sites, a few (fewer than one hundred worldwide) special sites should be selected for integrated studies. There should be a significant and integrated scientific commitment to these sites in the areas of hydrology, biogeochemical cycling, ecology, and satellite monitoring.

Facilities at these sites serve three purposes. First, they interface with operations at the next higher level of intensity (field campaigns and integrated ecosystem experiments) by providing starting points with infrastructure and background information sufficient to ensure high returns. Second, for facilities at the next lower level of intensity (long-term monitoring sites), studies at the higher-intensity sites will identify variables to monitor, define appropriate sampling intervals, and validate integrated models tailored to the specifics of each biome. Third, these facilities for high-intensity monitoring will be the primary barometers for changes in subtle parameters that affect aspects of ecosystem function without immediate or profound effects on structure.

It is anticipated that the energy, water, and carbon balance models to be tested using data sets from these sites will utilize meteorological data (e.g., temperature, humidity, precipitation, wind speed, and radiation) to provide the forcings and satellite data to provide the slowly changing surface boundary conditions (e.g., photosynthetic capacity and soil moisture estimates).

An ultimate goal for this measurement-modeling exercise is to develop the models that can be driven by the combined remote sensing and meteorological data sets to calculate fields of surface fluxes and associated forcings on the important ecosystem processes over the whole globe.

Accordingly, the proposed criteria for site selection are as follows:

- located within important biome centers or on biome transition zones,
- existing long-term research archive, and if possible, a paleoecological record (see chapter 3),
- nearby research institute for site support,
- "feasible" topography for flux measurement,
- presence of gauged or gaugeable watersheds of a reasonable size (5 to 20 km^2),
- away from excessive anthropogenic impacts, e.g., large-scale air pollution, and
- logistics: airfields nearby and road access.

To satisfy the modeling requirements, the stations would routinely acquire the following kinds of data:

- hydrological,
- meteorological (including radiation balance),
- ecosystem structure and productivity,
- land use and soil information,
- atmospheric optical depth,
- large-area surface fluxes (tower),
- trace gas concentrations, and occasionally fluxes, and
- selected satellite data.

Most of the data should be taken within a concentrated area of roughly 20 × 20 km, which allows a reasonable area for the sampling of satellite observations and airborne flux measurement. However, this core site should be located within a larger similar zone (100 × 100 km), which could be used for studies of the spatial and temporal variability of some of the parameters.

- Hydrological data. Various components of the hydrological cycle at local and regional scales are expected to respond to climatic shifts. Therefore studies of snow pack dynamics should include analyses of snowline retreat patterns over time and space. The timing of runoff originating from snow packs may be a significant indicator of general atmospheric warming at higher elevations.

Reservoir levels (minus consumptive use) can be taken as a long-term integrated measure of large-scale (regional) water surplus or deficit. Reservoir evaporation records also need to be examined as a measure of atmospheric evaporative demand. At present these activities may be undertaken in many areas, but it will be necessary to coordinate the data sets to provide larger regional analyses.

River discharge records, particularly from watersheds that have been monitored for long periods of time (>30 years), provide the basis for many hydrological investigations and need to be considered when locating the integrated monitoring sites. These records are valuable for model calibration and for determining the hydrological response of systems to climatic fluctuations. It will be necessary, however, to develop an understanding of the role of land use in the rainfall-runoff process of the watersheds to be examined so that these effects can be accounted for in assessing climatic impacts. Watershed-based studies will have the advantage of providing an integrated view of hydrological processes. Hydrological and other balances (e.g., nutrient) can be examined using the watershed as the unit of study. Furthermore, long-term (months to a year) evaporative losses can be estimated by resolving the various components of the watershed water balance and serve as a check on short-term monitoring approaches. The small (<150 km^2) watershed approach may be also be valuable for lumped soil moisture modeling studies, which can be integrated with remote-sensing-based procedures for estimating this variable.

Particular attention needs to be given to procedures for making large-scale estimates of soil moisture in a timely and cost-effective manner. In view of the critical role that soil moisture plays in controlling evaporative and discharge losses from watersheds, this monitoring effort should be regarded as a major priority. Therefore research into appropriate sampling schemes, new measuring devices (e.g., time domain reflectometry), and empirical prediction techniques will be required.

Since precipitation is a major determinant of moisture fields controlling biological activity and processes, the integrated monitoring sites should have carefully designed precipitation gauging networks.

- Data on ecosystem dynamics and biogeochemistry. Intensive monitoring at the sites should address ecosystem function at several levels, including (1) exchange of materials with the atmosphere, (2) exchange of energy with the atmosphere, (3) transport of materials to other ecosystems, (4) storage of biomass, nutrients, and energy, and (5) structural and functional species composition and diversity to all trophic levels. Since the biosphere observatories are intended to provide the most detailed long-term data available for any biome, exchanges of mass and energy should be monitored with an intensity sufficient to support accurate annual budgets of key elements (carbon, nitrogen, and phosphorus), atmospherically and biologically important chemical species (hydrogen oxide, carbon dioxide, ozone, methane, isoprene, nitrous oxide, nitric oxide, nitrogen oxides, and acidity), and forms of energy exchange (absorbed radiation, net radiation, sensible heat, latent heat, soil heat flux, and momentum transfer).

The appropriate technologies for monitoring these aspects of ecosystem function will vary somewhat from ecosystem to ecosystem but should be standardized as far as possible to allow interbiome comparisons. Exchanges with the atmosphere should be continuously measured with meteorological techniques. Eddy correlation is the most suitable existing technology for measuring fluxes of many of these components, although sensors for some (e.g., methane) need to be improved. The NCAR Atmosphere-Surface Turbulent Exchange Research (ASTER) facility provides one possible model for the kind of advanced eddy correlation systems that should be established in the sites. Other instruments like the Fourier Transformed Infrared (FTIR) spectroscopy combined with LIDAR also offer potential that should be explored.

Transport to other ecosystems should include hydrological as well as airborne transport. Rainwater and stream chemistry are critical components to monitor. The sites, as noted above, should include watersheds for which complete water and nutrient budgets can be obtained.

Storage monitoring should emphasize water, carbon above and below ground, and nutrients (nitrogen, phosphorus, potassium, calcium, and sulfur) in biomass and soil. Information on stored material is much more useful when it is accompanied by chemical data indicating resistance to decomposition (especially lignin content) and when it is partitioned into components of differing biological activity (e.g., below-ground carbon should be partitioned into root, microbial, and one or more soil fractions). Whenever possible, the storage terms should be measured directly and not estimated by difference from the flux studies.

The sites should also be positioned to serve as primary centers for monitoring diversity. The assessment and study of biodiversity should be based on regular censuses of all major taxa, including insects and microbes. In addition to an accounting of the number of species present and more sophisticated indices of evenness, the diversity monitoring should report structural diversity, including forest and soil stratification, and phenology.

• Satellite data. A goal for all of the studies on these sites is to enhance the use of satellite data for describing earth system processes, in the hope that the developed techniques can be applied globally over the sites. All available satellite data should be collated and archived together with ancillary observations, including photometric measurements, mesometeorological network observations and analyses, and radio soundings. These data sets should be continuously available to the scientific teams working in the area. A commitment must be made to regularly process subsets of these data to the point where they are directly useful to the scientific teams. This means that atmospheric and geometric corrections should be applied and in some cases the resulting fields processed into useful biophysical parameters (e.g., APAR).

Experiments

Experiments will be essential to

- understand the important processes controlling the interactions between the vegetated land surface, the atmosphere, and the hydrosphere, specifically, fluxes of energy, water, heat, carbon, and trace constituents;
- understand how these fluxes are associated with state variables associated with important ecosystems (e.g., community composition and physiognomy); and
- develop the links between the fluxes and state variables described above and those parameters amenable to remote sensing.

Two main kinds of experiments will be necessary to achieve these tasks: integrated ecosystem experiments and field campaigns.

Integrated Ecosystem Experiments

Integrated ecosystem experiments, manipulations that expose whole ecosystems to simulated climatic change or components of climatic change, will be critical for two reasons. First, they will provide the most concrete and complete evidence of the response of ecosystems to climatic change, and they can potentially provide it in a way that facilitates quantification of the mechanisms driving the responses. Second, they will provide the solid data for model development and testing.

Manipulative experiments will provide essential tools for the study of climatic change, but they will also be complex and expensive. These major efforts must be carefully designed to maximize their utility. Three principles can facilitate the choice of systems for study and the design of experiments. First, while there is a need to know the response of every ecosystem type to climatic change, it will not be possible to subject every ecosystem to one or more manipulative experiments. The likely time course of climatic change demands that we rapidly establish experiments in a broad range of ecosystems. Given the need for breadth, it is also important to emphasize ecosystems that will yield the greatest return on the resources invested. Important criteria to consider include the following:

- Experimental tractability. Ecosystems with small, short-lived dominates and rapid cycling of carbon and nutrients (e.g., grasslands) are inherently more accessible than ecosystems with large dominates and slow element cycling. It is critical that the ecosystems developed as model systems respond quickly enough that experiments proceed faster than climatic change. Rapid time responses and small stature also facilitate replication and factorial exposure to the full suite of components of climatic change, enhanced carbon dioxide, temperature, precipitation, and anthropogenic pollutants. In some ecosystems the suite of meaningful manipulations may include factors other than components of climatic change. Altering soil fertility, removing or augmenting populations at one trophic level, and changing light may be efficient ways to change ecosystem processes in a manner that contributes directly to understanding the mechanisms of responses to climatic change. Intensive study of carefully selected model systems, in combination with process studies (see the section "Fundamental Research and Laboratory Work" below) will greatly facilitate the design of the necessary experiments on ecosystems with long-lived, large-stature dominates and with slow carbon and nutrient cycling.

- Experimental relevance. It is critical that the ecosystems chosen for emphasis are subjected to experimental designs consistent with quantifying the mechanisms driving the responses to the manipulation. An emphasis on driving mechanisms is the most likely approach to making results of one experiment generalizable to a range of ecosystem types.
- Ecosystem sensitivity. Process studies and historical observation have identified several ecosystems likely to be greatly affected by or especially sensitive to climatic change (e.g., coastal marshes and arctic tundra, respectively). These systems merit high priority for study.
- Ecosystem importance to humans. Ecosystems that play a critical role in sustaining human populations, especially agroecosystems, deserve special priority for experimentation.

Second, natural and unintentional experiments provide excellent opportunities for addressing some questions. Some natural experiments can be based on the responses of ecosystems to unusual climatic conditions, like El Niño events. Others can utilize unique geological features, for example, sulfur dioxide and carbon dioxide from volcanic vents or carbon dioxide from natural springs. Areas of natural sea level subsidence may provide useful model systems. Unintentional experiments vary in scale from the entire globe to very local phenomena, for example, pollutant plumes from industrial sources to artificial range extensions from experimental agroforestry.

Third, the dynamics of climatic change and ecosystem responses warrant independent study. At this point, we have little information on the responses of ecosystems to the rate of climatic change. Most past experiments have exposed ecosystems to a step change in carbon dioxide, temperature, or precipitation. While this kind of manipulation may yield results comparable to gradual change, it may also cause different responses, and the differences may be ecosystem specific, depending on the time constants of the component processes. Study of these dynamics can be pursued with models, with experiments on model systems, and with process studies on the time constants of ecosystem components.

Field Campaigns

Field campaigns will be essential to

• study the biophysical controls exerted by the terrestrial vegetation on the transfer of energy, water, heat, and trace gases between the land surface and the atmosphere at appropriate time and space scales, and

• develop better methods for quantifying these controls and exchanges using satellite and standard meteorological data to permit global observation.

Again, the design, execution, and costs of these efforts will vary widely. Only their common elements will be discussed here:

• Monitoring studies. Field experiments should be sited within long-term monitoring sites, and measurement should be integrated with the long-term effort as far as possible.

• Theoretical framework. The design of these experiments should make allowance for both the correlative "top-down" and the biophysically based "bottom-up" approaches. While most researchers believe that the ultimately the bottom-up (reductionist) approach will provide the necessary information, it is clear that many remote sensing successes in the scientific arena are based on simple correlative techniques.

• Multisensor, multiangle, multiscale measurement strategies. It is almost certain that no single remote sensing measurement is going to provide the requisite information in isolation from other data. Likewise, to fully test theories and models, measurements have to be made simultaneously across several spatial scales, preferably using a nested sampling strategy.

• Redundancy and checks. Efforts should be made to allow independent estimates of key validation components, e.g., surface fluxes, using different measurement methodologies. Besides being a reasonable precaution, this may allow the validation of measurement technologies across different spatial scales. For example, a surface flux network could be used to test the validity of an airborne measurement technique, which could then be applied to a much wider area. Likewise, chemical analyses on outflows from whole watersheds could be compared with the aggregation of point samples taken within the area.

These large-scale field experiments are expensive, resource-intensive, and very demanding in terms of the necessary investment of scientists' time. As an example, the First ISLSCP Field Experiment (FIFE) (Sellers et al., 1988) involved the commitment of about 100 scientists, 50 support personnel, several million dollars worth of field equipment, and six aircraft.

However, these experiments offer unique opportunities for large-scale studies and are necessary for improving integration techniques for scaling from smaller to larger scales. In practice, it is hoped that the national science resource base will be able to support at least one such effort in the field per year without absorbing the energies of most of the interested community. The field experiments would be "nomadic," being focused on one integrated monitoring and process study site (see previous section) at a time and moving on to another site (different biome or different set of problems) on completion. As far as possible, the field campaigns should address both disciplinary and interdisciplinary problems. To the extent that ecological, hydrological, and biogeochemical processes are closely interlinked, this need to address a range of problems will be essential to the planning and execution of such experiments.

Field campaigns offer the best opportunity to check theories that extrapolate the results of leaf- or plant-scale phenomena up to the scales relevant to surface-atmosphere exchange and routine satellite remote sensing (i.e., several kilometers). The strategy developed in FIFE provides one template for this problem: surface studies are distributed within an area of around 20 km on a side that is also routinely observed using airborne eddy correlation, remote sensing equipment, and satellites. This design allows for the direct comparison of experimental results across most of the scales relevant to model development for GCMs (i.e., plant, site, and watershed scales) (see Figure 5.2). Experiments at much larger scales necessarily involve a dilution of the surface observation network and therefore cannot address the cross-scale problems directly. Large-scale experiments should be viewed as checks on model parameterizations at the scale of application (e.g., a GCM conformation of grid squares), about 1000 km on a side.

Field experiments also benefit the community by forcing scientists from different disciplines to collaborate in order to design, execute, and produce a concrete result. As catalysts for changing research policies and stimulating new interdisciplinary research, such experiments can be very useful tools to both the researchers and the agency personnel involved.

Simulated and Actual Impact Studies

A number of opportunities exist for conducting simulated or actual impact studies, including impacts of tropical deforestation, acid rain, and glacier retreat. Areas where there are large-scale ongoing impacts, e.g., the Amazon tropical forest, should have a high priority for combined long-term process studies and field campaign investigations. Such studies should address changes in surface energy and water balance, soil hydrological processes, soil physics and chemistry, species succession, and dynamics (socioeconomic) of land transformation. Where possible, paired sites should be used.

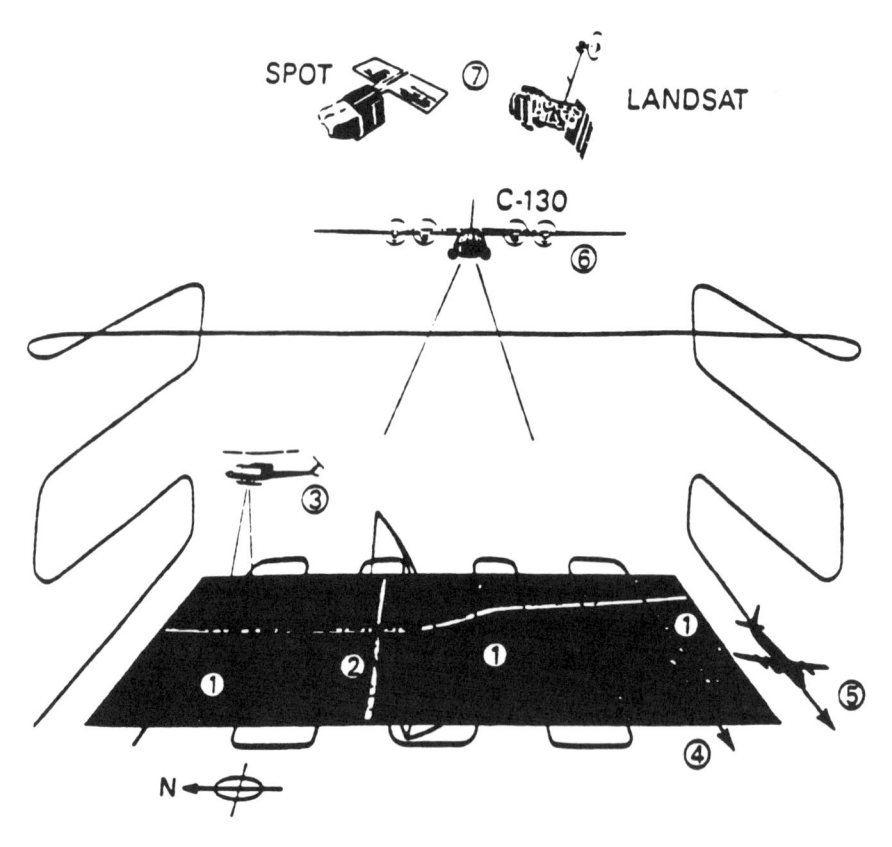

FIGURE 5.2 Satellite, airborne, and surface data acquisition at midday, August 4, 1989, in the middle of the FIFE-89 Intensive Field Campaign (Source: Sellers et al., 1990).

Fundamental Research and Laboratory Work

Fundamental research needs to be supported in a number of key areas, including remote sensing techniques, detailed ecological and hydrological process studies, and the development of improved measurement techniques.

Remote Sensing

Satellite remote sensing represents a feasible means of collecting appropriate, consistent, and temporally frequent data over the entire biosphere at reasonable cost. Its potential should therefore be exploited to the maximum extent possible.

In addition to the execution of field campaigns and the analysis of satellite data described in previous sections, research needs to be done to place remote sensing science on a more physical, less correlative basis. Model development and testing will be an integral part of this effort, and the development and testing of new instruments and sensor combinations will be essential. Particular emphasis should be placed on the following:

- Atmospheric, radiometric, geometric correction algorithms. Almost no satellite data are routinely corrected for atmospheric effects, partly because sufficiently robust algorithms are not available. Much more funding should be directed toward the development, selection, and implementation of such algorithms.
- Surface and spectral radiation interactions. To date, only correlative studies have been performed to relate surface spectral signatures with canopy properties (e.g., lignin/nitrogen ratios in foliage). More intensive modeling work (e.g., on foliar spectral properties) must be done to determine the links between plant biogeochemistry or ecosystem status and spectral signature.
- Microwave (active and passive) remote sensing. Microwave remote sensing offers some real promise in determining vegetation characteristics, soil moisture, and precipitation. More investment must be made in the development of instruments and theoretical treatments.
- Novel technologies. Atmospheric scientists have discussed the use of satellite-based or airborne lasers to determine aerosol optical properties, atmospheric boundary layer heights, and so on. The utility of such instruments and other novel sensors for surface and near-surface remote sensing should be investigated.

Ecological Process Studies

Process studies in ecology play two critical roles in global change research. First, they help identify the variables that need to be considered, set operational conditions, and simplify the design of integrated field experiments. Second, they provide a powerful, unambiguous arena for quantifying mechanisms and for testing the structure and parameterization of models that focus on component processes.

- Ecosystem processes. The challenge of developing ecosystem models competent to predict ecosystem responses to climatic change has several aspects that need to be supported with process studies. Some of these process studies (e.g., species change) fall within the traditional realm of population or community ecology, but others involve emergent properties at the ecosystem level (e.g., effects of climatic change on the species composition of the soil fauna and on the decomposition of existing organic matter) and indirect consequences of species effects (e.g., changes in the quality of soil organic matter resulting from changes in plant biome allocation or species composition).

- Ecophysiology. Four areas in plant ecophysiology clearly warrant increased priority. First, a better understanding is needed of certain processes that play a role in controlling plant responses to climatic change. For example, biomass allocation, phenology, and respiration are critical determinants of plant performance, but none is understood at a level suitable for developing a mechanistically based model.

Second, more work is needed on plant responses to simultaneous, interacting stresses and alterations in resource availability. Much of the past work in plant ecophysiology has focused on the response of one process (e.g., photosynthesis) to one stress (e.g., drought). In order to maximize the contribution to global change research, the focus should shift to the responses of multiple plant processes (e.g., photosynthesis, allocation, and nutrient uptake) to combinations of stresses and altered resources. Treatment factors should include biotic stresses like herbivory as well as abiotic factors like drought, altered temperature, and elevated carbon dioxide.

Third, plant ecophysiology is much better developed at the level of the leaf or the individual plant than at the level of the canopy or the ecosystem. Techniques for scaling mechanistic responses at the leaf level to canopy and ecosystem responses must be developed and evaluated.

Fourth, the last few years have seen advances in ecological models that predict plant processes (e.g., growth, photosynthesis, and biomass allocation) on the basis of resource availability rather than physiological or anatomical criteria. To the extent that they are generalizable, these models have the potential to dramatically simplify the task of predicting ecosystem responses to climatic change. On the other hand, it is also increasingly clear that potentially co-occurring plant species may differ in access to resources, efficiency of uptake, and efficiency of utilization. The interface between the indications of interspecific generality and species individuality is a critical area to resolve, both because of the potential simplification that results from a species-independent perspective and because the ecosystem consequences of species change are central components of ecosystem responses to climatic change.

- Community ecology. Community-level phenomena pose some of the greatest challenges for developing predictive models of the response of ecosystems to global change. Studies in community colony are replete with examples of singular events (e.g., arrival of a pest, pathogen, or predator) that dramatically alter the course of ecosystem development. The challenge for process studies is to assist in specifying the envelope of possible responses and quantifying the consequences of each for ecosystem development. It is especially important to relate ecosystem development to resource availability and biotic and abiotic stress.

- Population biology. Ecosystem responses to global change will almost certainly result from the combined effects of (1) plastic responses in existing genotypes, (2) migration of new genotypes, and (3) selection on new and existing genotypes. The critical challenge for population biology is to estimate the quantitative importance of each process, in a range of ecosystems.

Hydrology

Hydrological studies, particularly on scale effects, need to be supported aggressively. Some of this work can be supported under the field experiment umbrella (see the section "Experiments" above), but there is a need for fundamental research at the small scale. Important issues to be addressed are listed below.

- Soil physics and water relations. The enormous spatial variability in soil physics and microtopography makes a simple arithmetic averaging of surface characteristics for larger-scale applications almost meaningless. More experimental and theoretical work must be done to investigate small-scale variations within bulk treatment in a physically meaningful way. This may involve intensive work on small subcatchments coupled with distributed hydrological modeling efforts.
- Soil water and energy exchange. The flux of water from the soil to the atmosphere, whether through plants or the soil surface, is not well understood. A number of small-scale experiments should be conducted to address this issue. These experiments will necessarily involve ecological expertise and the application of scaling theorems as discussed above.
- Plant and water relations. Much progress has been made in the area of photosynthesis-transpiration relations, but ideally a great deal more needs to be done. Laboratory and small-scale field research represents the best avenues for initial progress.

Instrument Development

In addition to the remote sensing instruments described in the section "Remote Sensing" (above), there is a widespread need for improved instrumentation for both laboratory and field studies.

Efforts should be supported to improve existing techniques or develop new ones for the following kinds of measurements:

- Scalar and isotope concentrations. Here the need is for more measurements in addition to new methods.

- Eddy correlation, lidar sounding, and long-path-length measurements. All of these must be improved to allow more accurate and reliable estimation of surface fluxes.
- Precipitation, radiation, and so on. Methods for obtaining estimates of the large-scale values of these variables are essential for studies of the hydrological cycle and the surface-atmosphere energy balance.
- Hydrological state variables. Sensor development for estimating snow cover and depth and soil moisture distribution at scales ranging from in situ (local scale) techniques to satellite-borne instrumentation should be supported.

MODELING

Figure 5.3 shows the relationships between the (biophysical) land surface parameterizations (LSPs), the atmospheric general circulation models (GCMs), and the ecosystems dynamics models (EDMs).

Currently, there are few biophysically based LSPs in existence that have been successfully combined with GCMs (Dickinson and Henderson-Sellers, 1988; Sato et al., 1989). These models cover the relatively short-term biophysically controlled interactions between the vegetated surface and the atmosphere: radiative transfer (albedo), momentum transfer (roughness length),

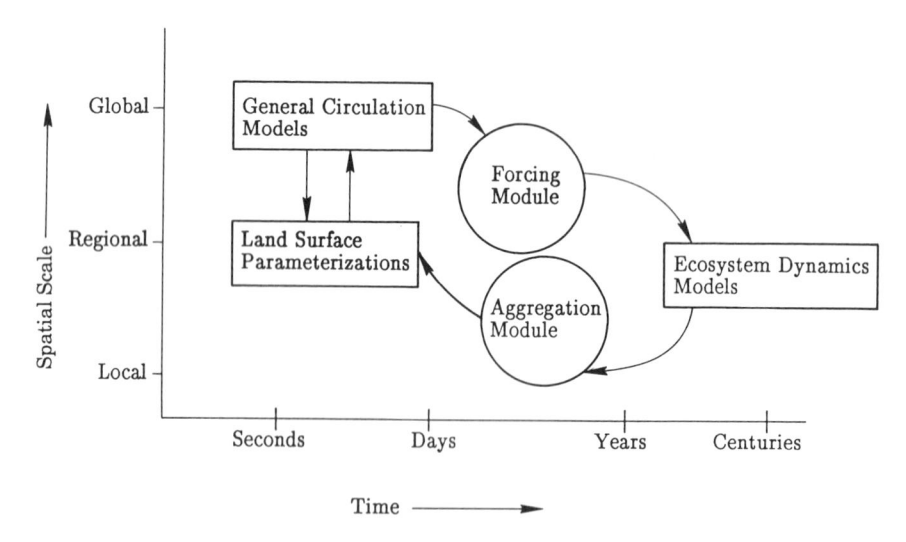

FIGURE 5.3 Relationships between land surface parameterizations, atmospheric general circulation models, and ecosystem dynamics models show disparity in time and space scales. For more effective global change research, two new "communication" techniques or "modules" must be developed to interpret results between model classes.

and exchanges of sensible and latent heat and carbon dioxide (biophysical control of evapotranspiration and photosynthesis). Essentially, the LSPs assume a static ecosystem structure and a prescribed phenology, which in turn define the albedo and roughness length characteristics of a given area and the evapotranspiration response as a function of soil moisture. Generally, the surface vegetation type, and hence albedo and roughness length, is prescribed from data, and the soil moisture field is initialized from offline climatological studies. As a result, these models in their current state have a limited utility for the study of global change because they merely represent an improvement over the abiotic "bucket" models described by Budyko (1974) and Eagleson (1982).

General circulation models of the atmosphere have improved considerably over 30 years of development to the point where they are the preferred tools for weather prediction and the study of climate. However, it is clear that these models will be subject to certain limitations for the foreseeable future. Most importantly, the models will be limited in terms of spatial resolution, representation of small-scale (subgrid scale) processes, and duration of run. It should also be remembered that each model possesses its own climatology, which differs from reality, and that an adequate description of the model climatology normally requires an extended series of runs. The normal GCM time step is on the order of 10 to 30 minutes (times much longer than this can lead to serious systematic error in the description of dynamical or physical processes), and this effectively limits the number or the length of runs that can be executed and analyzed by a research group.

Ecosystem dynamics models operate in an entirely different time and space domain from the LSP-GCM combinations, generally working on small spatial scales (meters to kilometers) and long time scales, integrating over centuries or millennia with time steps of up to one month. (For the time being, the discussion will exclude the "biogeographical" type of model, which describes the continental or global distribution of vegetation formations on the basis of "mean" climatology. These descriptions are not dynamic, as they operate as direct single-solution transforms of imposed climatic fields.)

It is likely that global change in the real world will affect elements of all three systems, as shown in Figure 5.4. The sequence of changes could be as follows:

1. A change in the physical climate system would bring about a direct biophysical response from the biota. For example, near-surface temperature or humidity changes would have a direct impact on photosynthesis and evapotranspiration rates (fluxes).

2. Changes in the surface biophysical response would directly affect the near-surface climate. The resulting feedback could be positive, neutral, or negative, depending on the circumstances.

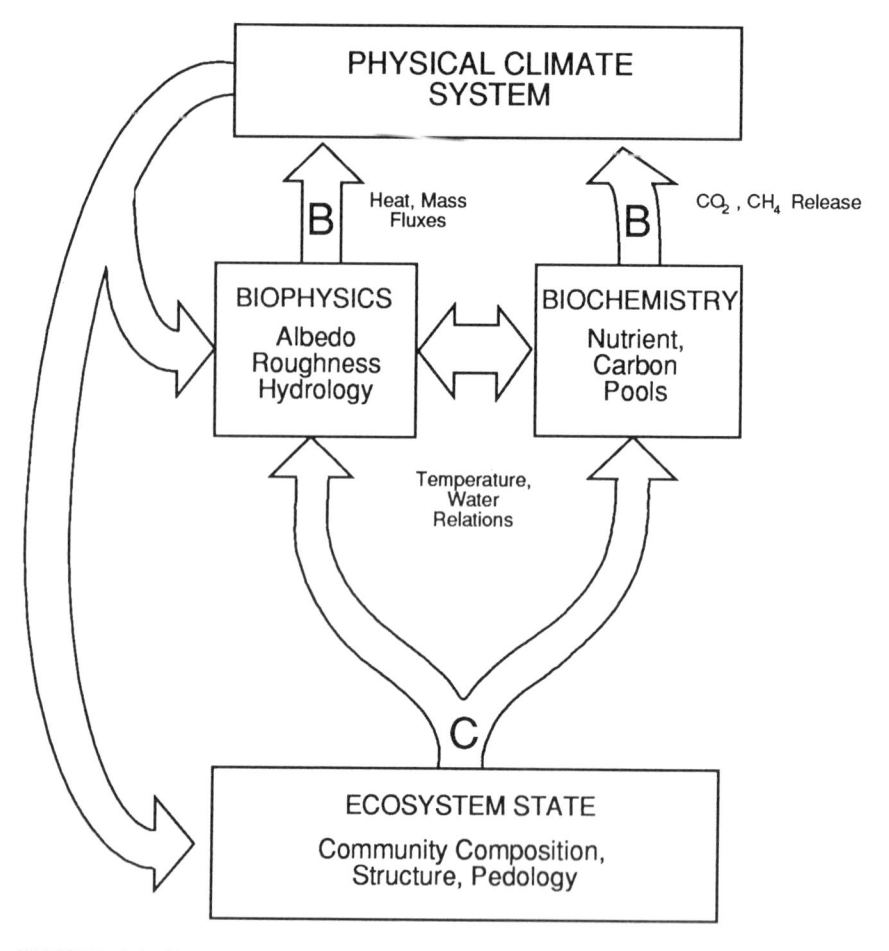

FIGURE 5.4 Important interactions between the vegetated land surface and the atmosphere with respect to global change. (a) Influence of changes in the physical climate system on the biophysical characteristics and ecology of the biome. (b) Changes in nutrient cycling rates; release of carbon dioxide and methane from soil carbon pool back to the atmosphere. (c) Ecological change in species composition results in changes in land surface characteristics of albedo, roughness, and soil moisture availability with possible feedbacks on near-surface climatology.

3. Changes in the near-surface climate or the surface biophysical response would have a direct impact on the forcing functions acting on ecosystem dynamics. Obviously, changes in ecosystem processes would be the first manifestation (e.g., rates of decomposition and mineralization), but these could be followed by changes in the gross structure (e.g., species composition and standing biomass). The degree, if any, of such changes is again highly variable and dependent on the locale and site history.

4. Changes in ecosystem structure and function can be expected to feed back onto the physical and possibly the chemical climate system.

As yet, the tools to investigate these phenomena have not been integrated in a scientifically defensible way. Although there has been some progress in coupling LSPs with GCMs, it is highly unlikely that either will be directly coupled with EDMs in the near future because of the gross disparity in time and space scales.

It can be convincingly argued that direct coupling is highly undesirable in any case. To force EDMs, many of which incorporate stochastic descriptions of ecological processes, good representations of the mean and variability of a region's climate must be applied: thus there is a need to repetitively apply many variations of a climatology to an EDM before a credible ensemble of results can be collected. In addition, the spatial scale of most EDMs is not consistent with that of GCMs: the answer, of course, is not to increase GCM spatial resolution to finer and finer scales, as this would result in problems similar to those discussed regarding temporal scales. For both of the above reasons it is clear that direct links between LSP-GCMs and EDMs are both impracticable and undesirable. However, before significant progress can be made in the area of medium- to long-term atmosphere-biosphere interactions, it will be necessary to construct more rigorous linkages between the LSP-GCMs and EDMs. It is proposed that this be done by constructing "forcing modules," to convert GCM output into "forcing" climatologies for EDMs, and "aggregation modules" to aggregate the effects of ecosystem dynamics changes into representative LSP parameter sets.

In spite of the general state of modeling described above, every effort should be made to support model development in every direction—GCMs, LSPs, and EDMs—as these represent the greatest opportunities for predicting the mechanisms and effects of global change. Some specific needs that merit special attention or that have not been addressed in previous reports are listed below.

Intermodel Transfer Packages

As discussed above, two kinds of intermodal transfer packages (ITPs) are needed—the first to allow communication from LSP-GCMs to EDMs and the second to allow communication in the other direction.

The first, the forcing module, would accept GCM output and generate the requisite "climate" for EDM applications. The module should take into account the following:

- Biases due to GCM climatology.
- Effects of GCM resolution and the parameterization of subgrid-scale processes.

- The likely range of microclimates produced by variations in topography, soil moisture, and pedological effects.
- The frequency and type of "extreme" events associated with a climatology in addition to the description of the mean condition.
- Results obtained from process studies and monitoring data sets.

Forcing modules would necessarily have to incorporate some background knowledge of the GCM's structure and performance. In this sense, they would be far more than simple extensions of quasi-stochastic "weather generators."

Aggregation modules would be used to analyze the results of EDM runs and would generate the requisite grid-scale parameters for LSP-GCMs. To a degree, the aggregation module is an inversion of the forcing module in that it attempts to generalize and integrate the specific and different results of EDM runs. The modules should take account of the following points:

- Integration of physiological characteristics probably cannot be done in a linear, arithmetic fashion. The "importance" of the contributions of different organisms to various fluxes, and so on, must be taken into account.
- The impact of spatially varying soil moisture should also be integrated to take account of nonlinear effects on the surface fluxes.
- Where appropriate, the effects of landscape pattern, e.g., repeating topographic units, should be integrated using ensemble averaging techniques.

Phenological Descriptions for LSPs

As discussed above, it is impracticable to place full EDMs within GCMs. However, some elements of vegetation phenology could be formalized and placed within LSPs. In particular, the following physiological phenomena should be described as functions of GCM prognostic variables (temperature, humidity, soil moisture, radiation, and so on): time series of green leaf area index; rooting depth; and roughness length, albedo, photosynthetic capacity, and maximum canopy conductance if these are not functions of leaf area index and rooting depth.

The models should be able to describe the seasonal course of vegetation attributes and provide a crude response to large interannual variations in precipitation.

Hydrological Models

Land hydrological modeling is currently split into two effectively noncommunicating camps:

- "Wet" hydrology, an extension of the traditional hydrology, which had its roots in engineering applications (e.g., channel routing and storm flow response). Many recent research efforts have been focused on small-

or regional-scale catchment models, with spatially distributed descriptions of rainfall interception, overland flow, and infiltration. Usually, these models have fairly simple evapotranspiration descriptions.

- "Green" hydrology, mainly concerned with the study of the biophysics of the evapotranspiration process using soil-plant-atmosphere models. The same basic models have been applied to describe processes on small-scale (agricultural) sites up to the scale of GCM grid areas.

Clearly some linkage between the different kinds of models is required. In particular, greater efforts must be expended to make the biophysical models into better descriptions of spatially heterogeneous surfaces. Also, some distillation of the wet hydrology models must be introduced into the LSPs: currently, all the LSPs use simple one-dimensional descriptions of the infiltration process with very simplistic representations of overland flow and spatial heterogeneity. The goal should be to provide good descriptions of total runoff losses as integrated over a month or more, rather than the extremely difficult objective of reproducing the correct timing of runoff losses.

Surface/Planetary Boundary Layer Models

Many of the problems in describing realistic feedback mechanisms between the surfaces and troposphere are associated with the description of mixing processes in the planetary boundary layer. Correct description of these is also important for some modeling inversion techniques driven by satellite remote sensing (e.g., determination of surface heat fluxes using meteorological and satellite data). Modeling efforts to address both goals should be encouraged.

Ecosystem Structure Models

The models discussed in previous sections can describe the effects of changes in land surface biophysical parameters (e.g., albedo, roughness, and moisture availability) and biogeochemical properties (e.g., nutrient cycling rates) as they are defined by a given ecosystem status (e.g., species composition, soil microecology, vegetation health, and leaf area index). Modeling techniques for describing ecosystem structures must be developed in parallel with the models discussed in previous sections. These models should address the issues of alteration of ecosystem structure due to changes in (1) the physical climate system, (2) atmospheric chemistry, and (3) land use change.

Depending on the intensity and type of change imposed in a given region, the ecosystem structure may be altered slightly (e.g., by adjustments in carbon dioxide exchange rates) or drastically (e.g., with changes in species composition or leaf area index). All types of changes may feed back onto the physical or chemical climate system (see Figure 5.4).

Prediction of terrestrial ecosystem change can be approached by using biogeographical methods as well as by using "full-up" ecosystem simulation models. It is strongly recommended that research efforts be encouraged on a broad front: the preliminary biogeographical models will provide us with the means to explore the possible sensitivity of land-surface- atmosphere interactions to changes in surface conditions. This approach is represented by Loop I in Figure 5.5, where a climatic change scenario leads to a simple definition of a new (steady state) distribution of biomes, which is then fed back through the climate modeling process to test for second-order effects due to the induced change in surface cover. Results from this kind of study will indicate which regions and biomes are important in terms of inducing second-order effects and thus merit further studies and interaction using kinetic ecosystem structure models (see Loops II and III in Figure 5.5). In this respect, none of the models should be regarded as an ultimate replacement for the rest; they have different roles depending on the level of detail required by either the climate modeling effort or the interests of biologists working on the effects of global change. In all cases the intermodel transfer packages discussed in the section above will have to be used as communication models.

Radiative Transfer/Plant Physiology Models

A number of modeling efforts have been partially successful in retrieving vegetation attributes from remotely sensed optical data. For the most part, these have concentrated on obtaining values for biometric properties such as leaf area index or biomass. More recently, efforts have been made to calculate physiological properties from observed radiances, including canopy (area-averaged) photosynthetic capacity and stomatal conductance. Research continues in the use of radar and passive microwave instruments for interpreting vegetation properties, but so far these efforts have focused on the retrieval of biometric properties and the classification of landscapes into different cover types.

The field of remote sensing of biospheric functioning is on the verge of providing invaluable information for the study of global change. The key to progress is clearly in the development of satellite data algorithms that can calculate appropriate states and rates associated with the terrestrial vegetation along with estimates of the uncertainty attached to each derived value. Such algorithms will have to address the following problems: satellite sensor calibration, sun-target-sensor geometry and the effect of atmospheric scattering, radiation transfer within the vegetation canopy and soil, and the

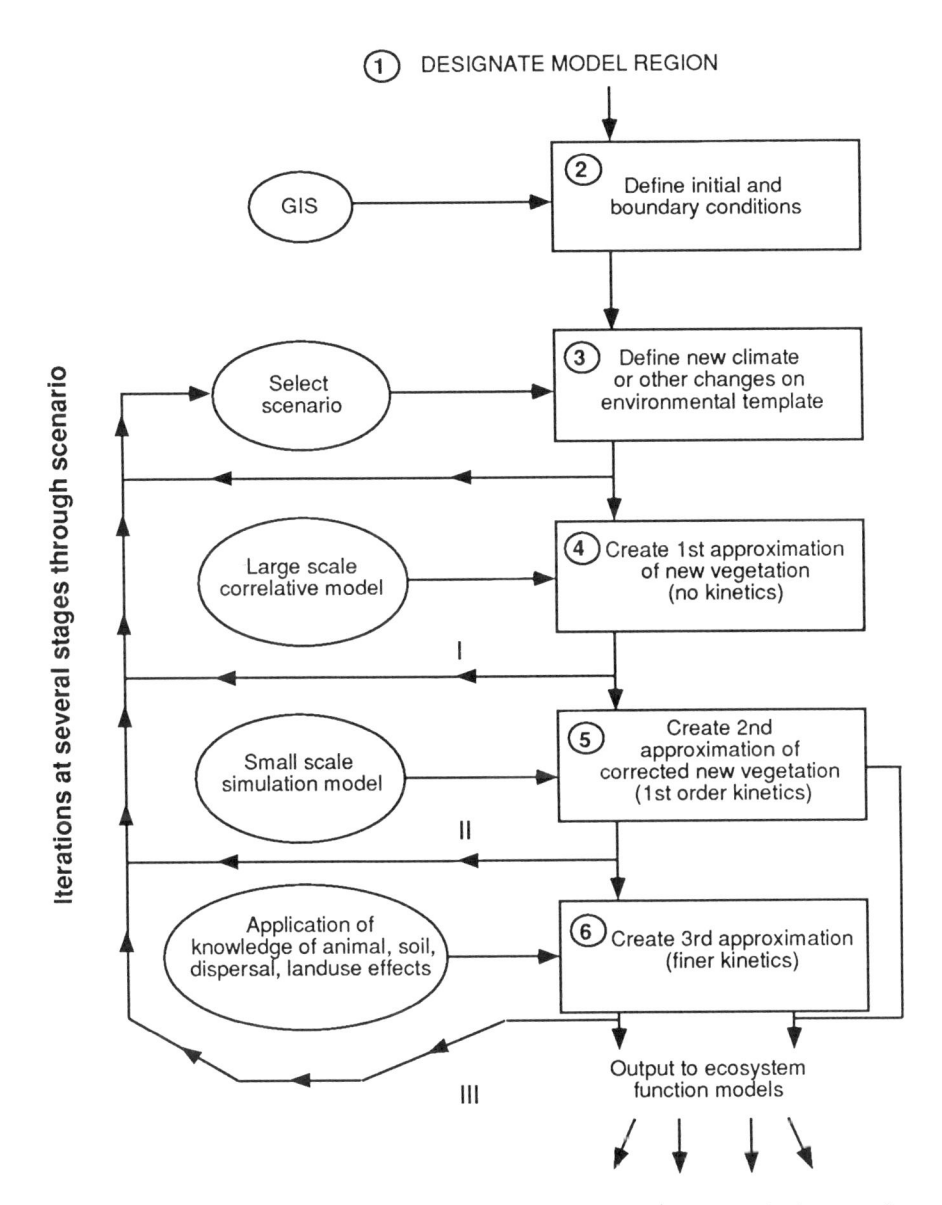

FIGURE 5.5 Proposed approach for modeling vegetation (ecosystem) change subject to large changes in forcing functions.

relationship between canopy scattering properties and relevant vegetation properties.

Although parts of the above problem can best be addressed by specialized researchers working in a loose confederation, the final goal of an integrated radiance-to-surface parameter algorithm must be kept firmly in mind. The achievement of the goal will require coordination of the talents of scientists working in several different disciplines: remote sensing technology, atmospheric physics, radiative transfer, plant physiology, and modeling. Ultimately, the effort could provide scientists with the means to calculate carbon, water, and energy fluxes over the land surface from satellite data.

Soil Genesis Models

Many of the models discussed above require some basic information about soil properties for their operation—soil physics properties in the case of GCMs and soil optical and nutrient properties in the case of the canopy radiative transfer/physiology models. Some of these properties can be derived using soil genesis models, which require data on the parent material, climatological regime, and vegetation cover as input. While these models cannot provide definitive production in most cases, they could have potential in terms of filling the gaps between reliable observations.

Sensitivity Analyses

All of the above sections have addressed the need to advance the realism and sophistication of different modeling efforts, in essence forming a "broad front" approach to the component parts of the issue of terrestrial biosphere-atmosphere interactions. Another important task must be coordinated with all of these—a sensitivity study on the effect of errors or uncertainties in the input data set of each model on the calculated product. Ultimately, it is hoped that all the individual models will provide products that can be used as input or validation for other models. For example, the products from the remote sensing algorithms could be used to prescribe surface conditions for GCM studies.

To make the best use of research resources, including money, time, equipment, and personnel, it is important that the sensitivity of each class of models to variations in their input parameter set be well understood. For example, if analyses indicate that GCM-simulated climates for continental interiors are sensitive to the successional stage of the vegetation cover there, it would be highly desirable to increase the flow of resources for improving the ecosystem dynamics models, which could then provide more realistic boundary conditions for the GCM. For this and similar problems, a gradualist approach to the sensitivity problem should be used. Simple schemes or parameter prescriptions should be used to determine the sensitivity of the model in question to variations in the input parameter set. The results of these basic tests should determine whether or not to invest heavily in more sophisticated approaches.

Summary

The above analysis has called out the need to vigorously promote research efforts in a few areas of obvious weakness. However, it should be reemphasized that modeling efforts in all the relevant areas should be supported to a much greater extent than they are now; these include the ecosystem dynamics models, land surface parameterizations, and atmospheric general circulation models. The descendants of current models will be the tools for understanding and predicting global change.

INFRASTRUCTURE

The modeling and data gathering tasks discussed in previous sections will not contribute to the overall goal of understanding and predicting global change unless there is a continuing effort to coordinate the activities. Defining the form of a governing coordinating body is beyond the scope of this document, but describing its purview is a necessity.

Operational Observations

There is no question that the array of operationally acquired meteorological, oceanographic, space-based, and other data is invaluable for earth system studies. However, most of the existing networks are not suitably configured for this work (e.g., aviation forecasting dominates many meteorological activities), and insufficient resources are dedicated to storing the data. All operational systems need to be considered as possible contributors to the earth system science effort, and a means of prioritizing and storing important data types needs to be formalized. Assembling a self-consistent long-term record of variables important for earth system science will require the implementation of new measurement networks, new information systems, and a high degree of collaboration among agencies.

Satellite Data Processing

The need to produce useful satellite data products for the scientific community has been emphasized above. A mechanism is needed to specify the list of desired products, with associated accuracy and precision requirements, and transmit this to the agencies so that resources, facilities, and personnel are dedicated to the task. This need is as urgent as the need for the development and launch of a new generation of instruments.

Centers for Research and Monitoring

Centers for research and monitoring should be the sites where more intensive, coordinated experiments and scientist training, as well as continuous monitoring-type observations, take place. An international effort should be made to establish such centers so that these research tasks are directly addressed. The centers should be sited in areas that are representative of a large and important vegetation formation, a "sensitive" area (e.g., a transition zone), or a benchmark area where there is a long research history and archive. To address the goals of ensuring a continuous, high-quality monitoring effort while suitable for intensive field experiments of the scale of FIFE or larger, there should be permanent research staff attached to each center. These staff, in cooperation with visiting scientists, should also carry out an educative and training function.

Education

As noted in other chapters, there is currently a critical shortage of trained researchers to carry out the task of earth system science research. A coordinated approach is required to recruit good students into the field and to train them to be able to participate in interdisciplinary research. This will take money, effort, and organization.

Interagency and International Coordination

There is a need to integrate the planning and implementation of measurement networks, modeling efforts, experiments, and education at the interagency and international levels. This requires the interlocking of experienced bureaucrats and practicing scientists.

Coordination among most large-scale experiments is usually fairly haphazard, and thus over-redundancy and gaps continue to plague their operational implementation. A central clearinghouse, or at least an information exchange, would be useful. Such clearinghouses lead to better coordination in the use of instruments and personnel.

REFERENCES

Antarctic Ozone Hole Special Issue. 1987. Geophys. Res. Lett. 13(12):1191-1362.
Broecker, W.S., and G.H. Denton. 1989. The role of ocean-atmosphere reorganizations in glacial cycles. Geochim. Cosmochim. Acta 53:2465-2501.

Budyko, M.I. 1974. Climate and Life. Academic Press. 508 pp.
Dickinson, R.E., and A. Henderson-Sellers. 1988. Modeling tropical deforestation: A study of GCM land-surface parameterizations. Quart. J. Roy. Meteorol. Soc. 114:439-462.
Eagleson, P.S. 1982. Land Surface Processes in Atmospheric General Circulation Models. Cambridge University Press. 560 pp.
Hansen, J., D. Johnson, A. Lacis, S. Lebedeff, P. Lee, D. Rind, and G. Russell. 1981. Climate impact of increasing atmospheric carbon dioxide. Science 213:957-966.
McElroy, M.B., and S.C. Wofsey. 1986. Tropical forests: Interactions with the atmosphere. Pp. 33-60 in G.T. Prance (ed.), Tropical Rain Forests and the World Atmosphere. Westview Press, Boulder, Colo.
Mooney, H.A., P.M. Vitousek, and P.A. Matson. 1987. Exchange of materials between terrestrial ecosystems and the atmosphere. Science 238:926-932.
National Research Council. 1984. Global Tropospheric Chemistry: A Plan for Action. National Academy Press, Washington, D.C.
National Research Council. 1988. Report on Global Change. National Academy Press, Washington, D.C.
Ramanathan, V. 1988. The greenhouse theory of climate change: A test by inadvertent global experiment. Science 240:293-299.
Rotty, R.M. 1983. Distribution of and changes in industrial carbon dioxide production. J. Geophys. Res. 88:1301-1308.
Sato, N., P.J. Sellers, D.A. Randall, E.K. Schneider, J. Shukla, J.L. Kinter III, Y.-T. Hou, and E. Albertazzi. 1989. Effects of implementing the simple biosphere model in a general circulation model. J. Atmos. Sci. 46(18):2757-2782.
Sellers, P.J., F.G. Hall, G. Asrar, D.E. Strebel, and R.E. Murphy. 1988. The First ISLSCP Field Experiment (FIFE). Bull. Am. Meteorol. Soc. 69(1):22-27.
Sellers, P.J., et al. 1990. Experiment, design, and operations. Pp. 1-5 in American Meteorological Society Symposium on FIFE, February 1990, Anaheim, Calif. American Meteorological Society, Boston, Mass.
Trabalka, J.R. (ed.). 1985. Atmospheric Carbon Dioxide and the Global Carbon Cycle. U.S. Department of Energy, Washington, D.C. (Available as NTIS DOE/E/R-0239 from National Technical Information Service, 5285 Port Royal Road, Springfield, VA 22161).

6
Terrestrial Trace Gas and Nutrient Fluxes

OVERVIEW

The composition of the global atmosphere is influenced strongly by the biosphere's activity. Although the importance of photosynthesis and respiration in controlling carbon dioxide and oxygen has long been known, the importance of biospheric processes controlling nitrogenous compounds such as nitrous oxide, nitric oxide, and ammonia, sulfur compounds such as hydrogen sulfide, and various hydrocarbons has only recently been appreciated. Moreover, human activities (industrial, agricultural, and others) now affect these natural biospheric processes to such an extent that they may, in many circumstances, overtake some of them in importance. For the first time in the history of the earth, these natural and human-caused atmospheric-biospheric processes may alter the global climate, with potential impacts on human welfare.

This chapter was prepared for the Committee on Global Change from the contributions of Paul Risser, University of New Mexico, Chair; Jim Brown, University of New Mexico; Stuart Chapin, University of California, Berkeley; David Coleman, University of Georgia; David Correll, Smithsonian Environmental Research Center; Mary Firestone, University of California, Berkeley; Robert Howarth, Cornell University; Daniel Jacob, Harvard University; Jerry Melillo, Marine Biological Laboratory; Robert Naiman, University of Minnesota; William Parton, Jr., Colorado State University; William Reiners, University of Wyoming; David Schimel, Colorado State University; Robert Sievers, University of Colorado; Richard Sparks, Illinois Natural History Survey; Jack Stanford, University of Montana; Peter Vitousek, Stanford University; and the National Research Council's Committee on Atmospheric Chemistry. Daniel Albritton, NOAA, participated as liaison representative from the Committee on Earth Sciences.

There have always been global changes caused by natural processes such as changes in solar activity, changes in the earth's orbit, volcanism, and plate tectonics. The global changes under consideration today, however, are affected by human activities and include a wide variety of causes and effects, such as stratospheric ozone depletion, tropospheric ozone formation, global warming and sea level change, drought, deforestation, desertification, and reduction in biological diversity. Climatic change has occurred in the past on many occasions, but the projected rates now are much faster owing to the combination of natural and human-caused processes. A major challenge is to distinguish between these natural and human-influenced changes and to predict their specific and cumulative impacts on the biosphere and its inhabitants.

Reliably predicting changes on the global scale of some of these processes requires an adequate understanding of the cycles of carbon, nitrogen, oxygen, sulfur, and phosphorus. These required understandings involve three important research components (CES, 1989):

- biogeochemical processes occurring within oceans and on the land,
- geophysical and biogeochemical processes that control the fluxes of compounds between the atmosphere and the aquatic and terrestrial biosphere, and
- meteorological and chemical processes that control the distribution and transformation of chemicals within the atmosphere.

Changes in patterns and rates of terrestrial biogeochemical cycling caused by both natural and anthropogenic processes can cause changes in the global atmosphere; for example, the increase in carbon dioxide and other trace gases in the atmosphere can alter global temperature and rainfall patterns. Conversely, global changes can influence biogeochemical cycling; for example, global warming can cause an increase in the release of carbon dioxide and methane from boreal forest and tundra soils. Thus the connections between the atmosphere and the terrestrial biosphere operate in both directions. The bidirectional relationships between the atmosphere and the biosphere, and the complexity of these interactions, are the subject of this chapter.

Problem Definition

Although these atmosphere-biosphere interactions are now recognized as extremely important for environmental changes at the global scale, the physical and biological processes that control the flux rates and magnitudes to and from many ecosystem types are inadequately understood. This lack of understanding is caused by the complexity of these biological and physiochemical systems, by the difficulty of measuring some of these flux exchanges in the field, and by the heretofore insufficient attention directed toward these crucial studies. Analytical methods must be developed for measuring trace gas and nutrient fluxes under ambient conditions, ecosystems need to be characterized in terms of the connections between nutrient pathways and trace gas sources and sinks, the physiological processes and biochemical controls of these processes need to be understood, and these processes must be well enough known to translate the results from local and regional scales to the global scale—and to predict their behavior under various conditions of global change.

General Approach

The general approach for the research initiative proposed in this chapter is designed to provide an adequate understanding of trace gas fluxes and reservoirs and of the flows of nutrients. The following are addressed in the chapter:

- Statement of the most crucial questions to be answered.
- Identification of the processes and variables that have the highest priority for attention.
- Description of the data that will be required to build and test algorithms for models describing and predicting these processes.
- Designation of the most appropriate geographical areas and the environmental conditions under which the studies should be conducted.
- Description of the experiments that must be conducted and the data that must be collected.
- Method for organizing the resulting data and information into coherent data sets and models for describing the processes and for predicting their behavior under alternative conditions of global change.

To proceed with this general approach, data and information must be provided from efforts discussed in other chapters of this report. Some of these interactions are shown in Figure 6.1. Arrows A and B refer, respectively, to the trace gases and nutrient fluxes that are essential components of the research programs proposed in this report. Conducting these studies will depend on (1) models and measurements describing chemical composition of and reactions in the atmosphere, (2) predictions of changing climate, (3) measurements of changes in land use, and (4) assessing the influence of other anthropogenic activities. Data and information about these input variables will be generated from other coordinated studies in the U.S. Global Change Research Program (USGCRP).

The key variables regulating the fluxes of trace gases to and from terrestrial ecosystems vary from gas to gas. Thus, measuring one set of variables for one gas may not be appropriate to the understanding of another gas. On

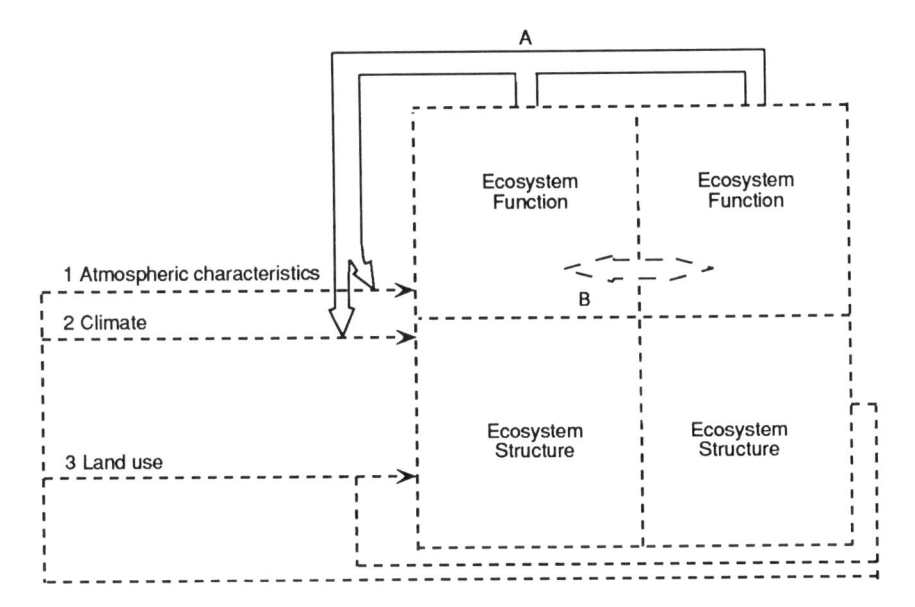

FIGURE 6.1 Coordination of the trace gas (A) and nutrient flux (B) studies with those described in other parts of the USGCRP (1, 2, and 3).

the basis of our current knowledge, it is possible to predict the necessary variables for modeling each gas. Table 6.1 is a first approximation of a summary of the variables needed to predict the exchanges of the major trace gases discussed in this report. The "influence" characteristics are expressed in general terms only. These variables may affect the fluxes through their influences on biomass loading, leaf resistances (e.g., stomatal opening), plant biological activity, soil chemical activity, microbial activity, or surface layer turbulence. Many of these variables can be mapped from satellite observations, others from land-based surveys.

This chapter consists of two related topics, namely, the exchange of radiatively, chemically, and biologically active trace gas species between the atmosphere and terrestrial ecosystems and the fluxes of nutrients within and among landscape units. Trace gas emissions lead directly to local effects but also may move laterally and affect adjacent or more distant landscape units. Materials (e.g., nutrients and pollutants) move within ecosystems but also move laterally in the hydrological cycle when they constitute or are attached to airborne particles. Moreover, lateral flows of nutrients, especially nitrogen and phosphorus, affect the sources and sinks of trace gases. The interactions between trace gas and nutrient fluxes are included in the described research programs.

The major global changes affecting the fluxes of water, sediment, nutri-

TABLE 6.1 Environmental Variables Regulating the Fluxes of Trace Gases from Terrestrial Ecosystems

Variable	Influence	Mapping Strategy
Surface temperature	Leaf resistance	Satellite
	Plant biological activity	Land-based
	Soil chemical activity	
	Microbial activity	
Solar radiation (PAR)	Leaf resistance	Satellite
	Plant biological activity	Land-based
Leaf area index	Biomass loading	Satellite
Greenness index (chlorophyll)	Leaf resistance	Satellite
	Plant biological activity	
Vegetation type	Leaf resistance	Satellite
	Plant biological activity	Land-based
Plant stress	Leaf resistance	Land-based
	Plant biological activity	
Surface roughness	Turbulence	Land-based
Sensible heat flux	Turbulence	Satellite
Surface wind	Turbulence	Land-based
Soil moisture	Soil chemical activity	Land-based
	Microbial activity	
Soil type	Soil chemical activity	Land-based
	Microbial activity	
Soil chemistry	Soil chemical activity	Land-based
	Microbial activity	

ents, and pollutants are land use and climate. Of these, altered land use will cause larger changes in these fluxes in the next years and few decades than will climatic change. However, climatic change will also affect lateral water flows and nutrient cycling, which, in turn, will affect trace gas flux and indirectly the climate. Thus there are specific links between climatic change, water and nutrient fluxes, and feedbacks to trace gas flux.

Many nutrient cycling studies to date have been conducted in relatively homogeneous areas (Likens et al., 1985). Much less is known about the transfer of nutrients across boundaries between ecosystems, but such transfers may greatly affect trace gas fluxes (Schimel et al., 1989). Therefore more careful attention should be given to these boundaries in terms of the fluxes that occur across boundaries and their controls.

At the global scale, the most significant issue concerning biogeochemistry is the effect of land use (e.g., cultivation and deforestation) on the flows of carbon, nitrogen, phosphorus, and sulfur from specific ecosystems and across the landscape. These losses from terrestrial systems occur by several atmospheric and soil surface and subsurface pathways and involve various interlocking biogeochemical cycles. Moreover, these exchanges ultimately affect the productivity and behavior of terrestrial, freshwater, and marine ecosystems.

The purpose of this chapter is to describe the crucial research questions, to identify the types of experiments to be conducted, to assess the availability of existing data, and to identify the locations and types of ecosystems that should receive the highest priority for immediate attention. As such, the recommendations are more specific than those found in the report of the Committee on Earth Sciences (1989), but less specific than some research plans, e.g., the International Global Atmospheric Chemistry (IGAC) program (Galbally, 1989).

The research needs discussed in this chapter are intended to be complementary to IGAC, a core project of the IGBP, and to address a crucial gap in understanding the fluxes of trace gases and materials to and from terrestrial systems. The focus of IGAC is principally on global atmospheric chemistry, with plans currently under development to include in the program the study of terrestrial sources of trace gases.

RESEARCH NEEDS

Trace Gases

Carbon Dioxide

Atmospheric carbon dioxide concentrations have been rising at 0.4 to 0.5 percent per year, apparently faster than ever before in the earth's history. Recently, they have increased even more rapidly. During the last decade, these increases have been associated with increasing amplitude of the annual cycle of atmospheric carbon dioxide and possibly surface air temperature of the earth. It is necessary to know the causes and effects of the accelerated rate of increase in atmospheric carbon dioxide, because it is a radiatively active greenhouse gas that has contributed to global warming and will continue to do so and because it has direct effects on ecosystems.

Research Priorities. The following research questions, listed in approximate order of priority, must be addressed to determine the causes and consequences of increasing atmospheric carbon dioxide. The priorities reflect the perceived importance of each research topic in reducing the uncertainty with which we can predict future changes in carbon dioxide. Chapter 7 addresses ocean-atmosphere interactions.

- How might climatic warming and associated changes in precipitation and nutrient status alter the biological carbon storage in ecosystems, especially those with large pools of stored soil carbon, through the redistribution of terrestrial ecosystems and through effects of enhanced carbon dioxide concentrations? Profitable approaches to carbon dynamics and global budget will be field and laboratory experiments, whole-ecosystem manipulations, and modeling, especially in tundra, boreal forest, and peat bog ecosystems, where there are large stores of organic carbon and where relatively large temperature changes may occur.

- How do increased atmospheric carbon dioxide and associated changes in moisture, temperature, and nutrients affect plant litter quality and the associated changes in soil respiration and nutrient mineralization? What are the effects of the resulting changes in nutrient availability on plant and microbial processes and on the sensitivity of intact ecosystems to enhanced carbon dioxide? This issue is best approached with a combination of field and laboratory experiments. These studies should be done in a range of ecosystems (e.g., wet versus dry and fertile versus infertile) where the strength of feedbacks between nutrient cycling, litter quality, and plant response to carbon dioxide might be expected to differ. The role of soil nutrient status is important in the context of global change, because industrial and agricultural pollution have dramatically increased the nitrogen availability of some ecosystems. Interactions of water availability and carbon dioxide fertilization must be studied, because projected climatic changes will involve changes in both parameters.

- How do ecosystem processes and different functional groups of plants (or specific key species) belonging to different ecosystems respond directly to changes in carbon dioxide and temperature in terms of rates of photosynthesis, allocation and net carbon balance, and indirectly in terms of competitive ability and such secondary processes as resistance to pathogens and herbivores? How are these carbon dioxide responses altered by interactions with other environmental stresses (e.g., drought, ozone, and nutrients)? Which ecosystems are the most sensitive? This research item differs from the item above in that it is plant-oriented rather than ecosystem-oriented and, as such, requires experiments at the level of individual plants.

- How would altered hydrological regimes predicted by global climatic models affect ecosystem carbon balance through changes in productivity and respiration in the short term and the characteristics of and the geographical distribution of ecosystem types in the long term? This issue must be approached through modeling and field and laboratory experiments.

- Why don't the perceived sources and sinks match the interhemispheric carbon dioxide studies? Much is known about the major sources and sinks for carbon dioxide and the global pattern of carbon dioxide transport in the atmosphere, but currently the perceived sources and sinks for carbon dioxide do not match the interhemispheric carbon dioxide gradient. In addition, the processes affecting trace gases (e.g., carbon dioxide, methane) in the paleoecological record must be reconciled with current understanding of carbon dioxide sources and sinks (see chapter 3). These issues should be addressed through the continuation and expansion of current programs to collect information that will most effectively differentiate among possible sources and sinks of carbon dioxide, e.g., enhanced plant growth and soil organic matter accumulation, fossil fuel burning, tropical burning associated with land clearing, and enhanced decomposition in boreal ecosystems. An expanded network for measuring atmospheric carbon dioxide and detailed isotopic measurements are needed to localize the major current sources and sinks for atmospheric carbon dioxide and to validate models that deal with the seasonal effects of terrestrial vegetation on atmospheric carbon dioxide.

• What are the consequences of landscape conversions, such as that of tropical forest to grassland, in terms of changes in stored soil carbon, evapotranspiration, energy balance, carbon balance, and nutrient status? Field measurements in appropriate ecosystems and modeling are needed to address this question.

• What is the effect of climatic change on episodic events such as fire frequency, the amount of carbon released, the resultant change in vegetation, and consequent changes in albedo, evapotranspiration, and plant production? How would the effects of human activities relate to those caused by climatic change? These effects should be addressed with satellite monitoring of disturbances such as fires and then related to surface moisture, temperature, and biomass. Patterns in natural and modified savannas, the boreal forest, and the tropics are of particular interest.

• What are the pools of biomass and soil carbon, net primary production, and ecosystem respiration in the world's ecosystems? All of the considerable field data need to be adequately collated and related to vegetation and soil maps for inclusion in global climate models. New data must be acquired by remote sensing of surface temperature, surface moisture, atmospheric water vapor concentration, and indicators of vegetation production and biomass.

All of these research questions address carbon balance at the ecosystem or global level and therefore are readily incorporated into ecosystem, regional, and global models. The major challenge will be designing models and experiments that link studies at the ecosystem level with inputs and predictions at regional and global levels (see chapter 5). It is important to recognize that as climate changes, so will the structure and species composition of these ecosystems. Thus models of ecosystems for today's circumstances will be inadequate for ecosystems for tomorrow, and therefore the models and analyses must be adaptable to changing ecosystem characteristics. The evolutionary approach described in chapter 2 is critical to success.

Methane

The concentration of methane is increasing in the atmosphere at a rate of about 1 percent per year and has approximately doubled in the past few hundred years. Methane is a greenhouse gas that, on a molecule-for-molecule basis, is about 20 times more effective than carbon dioxide in trapping heat. In addition to its role as a greenhouse gas, methane is an important sink for the hydroxyl radical in the atmosphere. The hydroxyl radical is the primary agent responsible for the oxidation and subsequent removal from the atmosphere of many reduced radiatively, chemically, and biologically important atmospheric gases. Depending on atmospheric nitrogen oxide concentrations and other chemical parameters, methane increases can change the atmospheric concentrations of the hydroxyl radical and hence change the atmospheric lifetimes and concentrations of several important gases, which would lengthen the time over which a species like methane contributes to radiative forcing of the climate system. Also, methane is an important source of water vapor in the stratosphere, and increases in stratospheric water vapor can have other significant global consequences.

Although the major sources of atmospheric methane are for the most part known, there is great uncertainty about the relative importance of these sources and which combination of sources and sinks is responsible for the rapid buildup of this gas in the atmosphere. The mechanism of methane production is fairly well known and results from anaerobic microbiological processes in wetlands, rice fields, and ruminants. Less well known are the environmental, physical, and biological processes that control the release of methane to the atmosphere. There are also major uncertainties about the anaerobic and aerobic sinks of methane in soils and sediments. One of the major questions that needs to be addressed is how changes in climate (e.g., warming in the northern high latitudes) may affect the global methane cycle. Methane sources in the tundra and wetland regions of the subarctic are major natural sources of atmospheric methane. A better understanding of ecosystem processes that control methane fluxes to and from the atmosphere and the impact that environmental changes may have on these processes is required if reasonable predictions of future atmospheric concentrations of methane are to be made. In addition, much of the methane in the high-latitude north is sequestered in permafrost and sediments as clathrates, which could serve as very significant sources of atmospheric methane if warming occurs. Similarly, a rise in sea level or warming of the oceans could also release marine clathrates. While this methane sink has been identified, its characterization, e.g., the temperature dependencies of the chemical reactions, needs to be better quantified. Significant research is in progress on methane, and new studies should be coordinated with projects proposed by NASA, the International Global Atmospheric Chemistry Program, and IGBP.

Research Priorities. The research activities required to better define biogeochemical budgets and cycling of methane were detailed in a recent Dahlem conference (Schimel et al., 1989). The major research activities required under the USGCRP are as follows:

• Process studies that relate methane production, consumption, and fluxes to environmental parameters, to human activities such as burning and livestock farming, and to changes in ecosystem structure and function need to be conducted for ecosystems of known or potential methane sources (e.g., wetlands, tundra, rice agriculture, and landfills). The study of the response of high-latitude northern ecosystems to environmental change should be studied through large-scale manipulative field experiments. In major rice-growing areas (e.g., India and China) the effect of cultivation practices on methane production, destruction, and atmospheric fluxes requires attention. Also, atmospheric pollutants could affect trace gas fluxes. Thus process studies must include interactions with pollutants.

• Improved instrumentation for the direct measurement of methane fluxes over small- and large-scale regions must be developed in order to improve our understanding of the relationship between fluxes and ecosystem processes and dynamics. Emphasis should be placed on the integration, or scaling, of information obtained from simultaneous chamber, tower, and aircraft flux measurements.

• Better spatial and temporal coverage of atmospheric methane concentrations and isotopic composition (carbon and hydrogen) in source regions must be obtained in order to apportion the global sources of atmospheric methane and understand the ecological and environmental controls of methane releases to the atmosphere. More extensive studies of isotopic composition of methane as a function of source and production and destruction processes should be made in order to use atmospheric isotopic information to better understand the biogeochemical budgets and cycle of methane. With such data it will be possible to more accurately model and determine the regional fluxes of methane to the atmosphere.

• In order to fully understand the atmospheric methane cycle, improved estimates of the atmospheric oxidation by the hydroxyl radical must be obtained. This requires a more complete understanding of atmospheric photochemistry than is currently available. Specifically, it is necessary to either directly or indirectly determine the concentration of hydroxyl radical in the atmosphere and the chemical processes that control this concentration. Thus it will be necessary to determine if a portion of the methane increase results from a reduction in hydroxyl radical concentrations (and hence a diminished oxidizing capacity of the atmosphere) through increases in the atmospheric concentrations of hydroxyl radical sinks, e.g., methane, carbon monoxide, and volatile organic compounds.

Volatile Organic Compounds

Depending on oxides of nitrogen concentrations, a number of volatile organic compounds (VOCs) are photochemical sources or sinks of tropospheric ozone—a toxic gas and a greenhouse gas—and as such they may play a significant role in global warming (see the section "Tropospheric Ozone" below). In addition, VOCs compete for oxidation by the hydroxyl radical with other atmospheric species, in particular methane; changes in the VOC budget could therefore affect the methane budget. The importance of the terrestrial biosphere as a source of VOCs is still poorly understood (Logan, 1985). Only a fraction of the large number of biogenic VOCs have been identified in the atmosphere, and few data on emission rates are available.

To identify and quantify the role of VOCs in atmospheric processes, it will be necessary to establish a comprehensive inventory of biogenic VOCs in the atmosphere, their emission rates from different types of ecosystems, and the environmental variables determining these emission rates. Additional studies of the chemistry of biogenic VOCs need to be made in the laboratory.

Research Priorities. The following research needs are listed in order of priority:

• Accurate techniques for identifying and measuring individual and cumulative VOCs fluxes and atmospheric concentrations (down to the pptv range) must be developed. A top priority should be to develop analytical instrumentation that can be operated from aircraft or better sampling and preconcentration techniques, since concentrations seem to be affected by sample storage. Once such instrumentation is available, large-scale field studies of atmospheric concentrations should be conducted to determine the regional and continental budgets of biogenic VOCs. Particular focus should be placed on tropical and mid-latitude forests, as biogenic VOCs may be strong modifiers of atmospheric photochemistry over these regions (Logan, 1985; Tingey et al., 1979).

• Improved measurements of fluxes are needed. The two methods currently used are (1) branch enclosure measurements and (2) inversion of measured atmospheric concentrations using chemistry-transport models. These methods have provided valuable information, but they are not fully satisfactory. Branch enclosure measurements are intrusive, and the resulting emission data will be biased to the degree that the biological functioning of the enclosed branch is impaired. Use of chemistry-transport models suffers from our poor understanding of the atmospheric reactivities of biogenic VOCs and of their decomposition products. Development of fast-response instrumentation for measuring atmospheric concentrations of biogenic VOCs is a top priority; such instrumentation would allow direct, nonintrusive measurements of fluxes by the eddy correlation technique.

• The sensitivities of biogenic VOC emissions to environmental factors need to be determined in the field and in the laboratory. Most data available so far are measurements of the effect of light and temperature on the emissions of isoprene and pinenes, for a few plant species (e.g., Tingey et al., 1979; Yokouchi and Ambe, 1984). Relatively few data are available on the effects of other potentially important factors such as water stress, air pollution stress, and fertilization. Laboratory studies should be aimed at increasing our understanding of the fundamental biotic mechanisms controlling the emissions of biogenic VOCs by various types of vegetation.

• The atmospheric reactivities of additional biogenic VOCs eventually need to be determined in the laboratory. Rate measurements of the reactions with the hydroxyl radical, ozone, and nitrogen trioxide made to date have allowed an assessment of the potential of specific VOCs to play a significant role in atmospheric chemistry. Reaction mechanisms for species found to be important should be investigated in detail, with a focus on the fate of the decomposition products (particularly the short-lived organic radicals). Environmental chamber experiments would provide a first assessment of the potential of biogenic VOCs as photochemical precursors of ozone and carbon monoxide, and as sinks for the hydroxyl radical and ozone. More precise kinetic investigations should also be conducted to understand the fundamental chemical mechanisms involved in the photochemical decomposition of biogenic VOCs.

Sulfur

Soils and terrestrial plants are known to emit a number of reduced sulfur species to the atmosphere including dimethylsulfide (DMS), hydrogen sulfide (H_2S), carbonylsulfide (COS), carbon disulfide (CS_2), and methyl mercaptan (CH_3SH). Coastal salt marshes and wet tropical soil might be significant sources of H_2S (Delmas and Servant, 1983; Andreae, 1985; Andreae et al., 1988; 1989). The available data indicate that these terrestrial emissions are in general much weaker than biogenic emissions from the oceans and are dwarfed by sulfur emissions from anthropogenic sources. Emissions from salt marshes could account for a significant portion of the global atmospheric sulfur budget, despite the small area involved (Andreae, 1985). Emissions from tropical soils and vegetation appear to be responsible for the background concentrations of sulfate observed over the Amazon Basin (Andreae

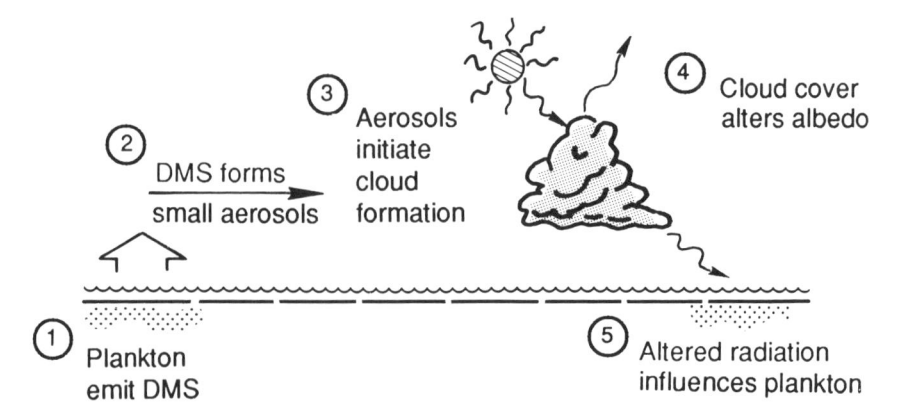

FIGURE 6.2 Proposed relationship between oceanic plankton and cloud cover.

et al., 1990) and could thus regulate cloud structure over tropical forests under pristine atmospheric conditions.

The terrestrial biosphere could be a factor for climate regulation through its effect on the atmospheric budget of COS. COS is the longest-lived sulfur species in the atmosphere; it can be transported to the stratosphere, where it provides a source of stratospheric sulfate, thereby affecting planetary albedo (Figure 6.2). Preliminary studies have indicated rapid uptake of COS at the stomata of plants, suggesting that vegetation could provide the major global sink for COS (Goldan et al., 1988).

The topic of acid deposition is discussed in the section "Nitrous Oxide and Reactive Nitrogen Compounds" (below).

Research Priorities. The following research needs are listed in order of priority.

- Measurements of biogenic sulfur emissions are needed from many more types of terrestrial ecosystems. Measurements of H_2S, DMS, CS_2, and CH_3SH should focus on regions thought to be potentially significant regional and global sources: wet tropical regions, coastal marshes, boreal forest peatlands, and tundra bogs that could be affected by changes in permafrost; anaerobic environments such as rice paddies and landfills; and industrial sources. The global distribution of terrestrial biological sinks and sources of COS needs to be quantified.
- Environmental variables affecting biogenic sulfur emissions need to be better understood. Preliminary studies suggest that vegetative emissions depend on temperature and insolation (Andreae et al., 1989), but the underlying mechanisms are unknown. Particular focus should be placed on un-

derstanding the factors that regulate H_2S emissions from salt marshes, as climatic change could alter dramatically the global surface area occupied by these ecosystems.

- Possible increases in reduced sulfur emissions to the atmosphere as a function of nutrient and sulfate inputs need to be investigated, particularly for the above ecosystems.
- Biomass burning has been proposed as a major source of atmospheric sulfur over tropical regions (Andreae et al., 1988), but very few data are available. Aircraft and land-based studies are needed to document the possible importance of this source.
- The role of terrestrial ecosystems as a sink for atmospheric COS needs to be investigated further. No data are available for the tropical forests, where vegetative uptake of trace gases in general appears to be particularly efficient.
- At night, plants appear to constitute net sinks for H_2S and DMS; research is needed to understand the mechanisms for this uptake.

Tropospheric Ozone

Concentrations of tropospheric ozone in the northern hemisphere appear to have risen steadily over the past few decades (Logan, 1985), and its photochemistry is the major source of the hydroxyl radical near the earth's surface. Tropospheric ozone—a greenhouse gas—is produced within the troposphere by oxidation of carbon monoxide and VOCs in the presence of nitrogen oxide. Ozone also enters the troposphere from the stratosphere. It is removed by photolysis, chemical reactions, and deposition to the earth's surface. The observed increase of ozone concentrations in the northern hemisphere is generally attributed to anthropogenic emissions of nitrous oxide, carbon monoxide, and hydrocarbons. New studies are required, but efforts are under way in the International Global Atmospheric Chemistry and the NASA Earth Observation System programs.

Biosphere-atmosphere interactions may play an important role in the global climate and budget of tropospheric ozone. First, deposition through the stomata and on the cuticles of plants is thought to provide a major sink. Second, interactions of biogenic VOCs with nitrous oxide of anthropogenic origin could elevate ozone production substantially over preindustrial levels, thus contributing to the rise in ozone concentrations. Much has been learned, however neither of these processes is thoroughly understood, and there is a serious need for further research. Finally, ozone injury to vegetation may alter the function of ecosystems, and eventually their structure, with possible feedbacks on climate.

Research Priorities. The following studies include those of the highest priorities.

- The highest priority is to improve the data base for ozone deposition to various terrestrial ecosystems. Eddy correlation measurements from towers and aircraft are delicate but provide at this time the best means to collect such data. Measurements from towers are particularly useful as they allow simultaneous monitoring of the environmental factors likely to influence ozone deposition fluxes. These factors include micrometeorological variables (e.g., heat flux, humidity, temperature, friction velocity, and light intensity within the canopy) and parameters of ecosystem structure and function (e.g., leaf area index, surface roughness, and stomatal and cuticular resistances). Models should be designed to relate the measured ozone fluxes to fundamental meteorological and ecosystem properties (cf. Meyers and Baldocchi, 1988).
- Tropospheric ozone is variable in distribution, especially near anthropogenic sources. Ozone affects the growth of vegetation directly and indirectly affects the ability of plants to respond to secondary influences such as drought, insects, and other pollutants. It is necessary not only to generate ozone concentration and flux data but also to understand the effects of ozone on vegetation, especially in relation to other environmental conditions.
- The physiological and chemical processes controlling ozone uptake, the release of other chemicals by plants stressed by ozone exposure, and the responses of various vegetation types to ozone exposure should be investigated in the laboratory and field. At the level of individual plants, data are needed for the rates of reaction of ozone at the plant mesophyll and at the plant cuticle. If most of the ozone uptake by plants takes place at the stomata, then changes in ecosystem function could have important implications for ozone deposition. An interesting issue, particularly in light of the effects of enhanced carbon dioxide on stomatal closure, is the possibility of stomatal closure due to ozone stress. Such an effect would introduce a positive feedback to the rise in tropospheric ozone levels. On the other hand, chamber exposure to whole plants and segments of ecosystems demonstrates significant responses to elevated ozone levels. The generality of this process must be evaluated in both urban areas and more remote areas that are influenced by human activities.
- Large-scale field measurements of atmospheric composition should be conducted to evaluate the contributions of biogenic VOCs and nitrous oxide to ozone production (see NRC, 1984; Lenschow and Hicks, 1989). As pointed out in the section "Volatile Organic Compounds" above, biogenic VOCs could be important photochemical precursors or sinks of tropospheric ozone, although more data are needed to determine their concentrations and reactivities in the atmosphere. Long-range transport-photochemistry mod-

els should be developed to interpret the observations of atmospheric composition in terms of ozone production rates and estimate the contributions from biogenic emissions to these production rates.

Carbon Monoxide

Carbon monoxide is emitted directly to the atmosphere by human activity (e.g., fossil fuel burning and biomass burning) and is also produced by atmospheric oxidation of hydrocarbons. Emission of carbon monoxide by the biosphere may be globally important, but the data base is very limited. Carbon monoxide has an atmospheric lifetime of a few months against oxidation by the hydroxyl radical, its principal sink.

There is strong evidence that atmospheric concentrations of carbon monoxide are increasing in the northern hemisphere as a result of anthropogenic emissions. Because of competition between carbon monoxide and methane for oxidation by the hydroxyl radical, a rise in carbon monoxide could have consequences for a parallel rise in methane. Also, carbon monoxide is a photochemical precursor of ozone, so that enhanced production of tropospheric ozone could follow from higher carbon monoxide concentrations (Logan et al., 1981). Thus increases in carbon monoxide concentration can contribute to global warming by causing increases in the atmospheric concentrations of two major greenhouse gases.

Only a few measurements of biosphere-atmosphere exchange of carbon monoxide have been reported in the literature. It appears that soils can be both sources and sinks of carbon monoxide, the net direction of the flux depending strongly on the environmental conditions (Conrad and Seiler, 1985). Some preliminary measurements suggest that plants are a significant source of carbon monoxide (Seiler, 1978), but the research is incomplete. Measured atmospheric concentrations of carbon monoxide in tropical forests of Africa and South America indicate evidence for a strong direct natural source of carbon monoxide (Marenco and Delaunay, 1985). Oxidation of biogenic VOCs constitutes another potentially important natural source of carbon monoxide (Hanst et al., 1980).

Research Priorities. The available literature does not permit an assessment of whether emissions of carbon monoxide from soils and plants could significantly affect the carbon monoxide budget on a global scale. There is a need for exploratory research aimed at addressing this issue. The geographical data base for carbon monoxide natural emissions should be expanded, particularly in tropical regions. Fast-response instrumentation for measuring atmospheric carbon monoxide concentrations is now available, so that eddy correlation flux measurements can be conducted from towers and from aircraft. Regions where biogenic fluxes of carbon monoxide can make a significant contribution to the atmospheric budget must be identified, and the environmental factors affecting carbon monoxide emission in these regions must be examined. Finally, fundamental laboratory studies are needed to understand the biotic and abiotic processes regulating the biogenic flux of carbon monoxide.

Nitrous Oxide and Reactive Nitrogen Compounds

The release and uptake of nitrogen-containing trace gases by ecosystems have important implications for atmospheric composition and nutrient fluxes and cycling. Nitrous oxide, a significant greenhouse gas with a long atmospheric lifetime, is the most important agent in natural ozone destruction in the stratosphere and has been increasing in atmospheric concentration at the rate of about 0.25 percent per year. Nitrous oxide is formed in soils by both nitrification and denitrification processes in both natural and agricultural ecosystems. Nitric oxide, also produced in soil and released to the atmosphere through nitrogen cycling, is a chemically reactive gas that regulates tropospheric ozone production. Nitric oxide is also released to the atmosphere in significant quantities through the burning of biomass. Ammonia is released to the atmosphere from plants, fertilized soils, and animal wastes.

Ammonia is the primary basic gas in the atmosphere and can play a major role in controlling the acidity of precipitation. Ammonia can be a significant vector for the medium-range transport of nitrogen into and from ecosystems. Nitric acid is produced in the atmosphere through the oxidation of nitric oxide and nitrogen dioxide and is rapidly deposited through dry and wet deposition to the surface of the earth. Deposition of nitric acid (and of nitrate salts resulting from the neutralization of nitric acid) can be an important nutrient input to ecosystems and, since nitric acid is strong, can also contribute to ecosystem stress. The input of nitric acid and its nitrate salts (as well as sulfur oxides near sulfur sources) can be especially important to ecosystems near regions where anthropogenic nitrogen oxide emissions are large (e.g., northeastern United States, Central Europe). Secondary alkyl nitrates and peroxyacetylnitrate (PAN) are formed in the atmosphere and can transport nitrogen from urban to rural areas, affecting ozone levels. It is necessary to obtain an improved understanding of how the budgets and cycling of nitrogen within ecosystems are connected to the atmospheric fluxes of nitrogen-containing gases.

Research Priorities. The following priorities are identified:

• Because of the important role of nitrous oxide as a greenhouse gas and the large uncertainties about sources, increasing our understanding of the ecological and environmental factors that control the atmospheric source strength of this gas for different geographical locations is a high priority.

The NSF Long-Term Ecological Research Program and the proposed IGBP Regional Research Centers and the Tropical Soils Biology and Fertility (TSBF) programs provide a framework for such studies. These studies will require direct flux measurements of nitrous oxide from different ecosystems over a wide range of ecological and environmental conditions. Tropical areas should initially receive a high priority for such studies, since these areas, which are believed to be important nitrous oxide source regions, are undergoing rapid changes in land use practices that are expected to significantly alter production and the fluxes of nitrous oxide to the atmosphere.

• The topic of acid precipitation, including both nitrogen and sulfur compounds, requires significant additional study. In particular, there needs to be a far better understanding of the conditions under which this process results in fertilization of ecosystems and of the conditions that result in a toxicity from acid precipitation itself or from other pollutants. Moreover, the research should not only aim to understand the effects of acid precipitation on the biosphere and the consequences for trace gas and nutrient fluxes, but also study the secondary consequences such as the release of aluminum and other materials from soils receiving acid precipitation.

• Improved understanding about the amount of nitrous oxide produced through the fertilization of agricultural systems must be developed through process studies designed to understand the mechanisms relating nitrous oxide and ammonia fluxes and the type of fertilizer, agricultural practices, application method, crop structure, soil type, and prevailing climate.

• Mechanistic ecosystem models that relate nitrous oxide fluxes to soil microbiology, micrometeorology, soil type, and environmental conditions must be developed, and the processes better understood.

• The global distribution of biological nitric oxide and ammonia fluxes to the atmosphere needs to be better established, and the importance of these emissions in atmospheric photochemistry and precipitation chemistry defined.

• Improved fast-response instrumentation for direct determination of fluxes by micrometeorological techniques of all nitrogen-containing trace gases over both small and large spatial scales must be developed. Such instrumentation will permit a more detailed assessment of the biological source strengths of atmospheric nitrous oxide, ammonia, and nitric oxide and allow a more accurate determination of the atmospheric input of nitrogen to selected ecological regimes.

• The uptake of nitrogen oxides and ammonia by terrestrial plants needs to be investigated. Preliminary studies suggest that this uptake could provide a significant sink for several gas species in some regions. For example, trees in the Netherlands are known to be major sinks for ammonia released by agricultural operations, and such a process could be important in other regions as well. Other recent studies have suggested that vegetation can provide a sink for nitrogen dioxide, which would inhibit the export of soil-derived nitric oxide to the atmosphere.

Nutrient and Material Fluxes

Fluxes Across Terrestrial Systems

Under current natural and human-influenced conditions, massive amounts of sediment, nutrients (including fertilizers), and pollutants are transferred across terrestrial portions of the biosphere and into streams and eventually the oceans. The increase in transport due to human activity above natural rates has assumed major proportions with global biogeochemical cycles. These transfers are caused by various land uses and are affected by changes in global climate, and, in turn, these fluxes affect global climate by influencing hydrology, trace gas exchange, and other processes. A basic research issue is to understand the flows of water, sediment, nutrients, and pollutants across terrestrial systems and into the air or aquatic systems and the responses of these systems. We must determine how global climatic change will affect these fluxes, especially the reciprocal interactions between the hydrosphere and the biosphere. To accomplish this objective, the fluxes of these materials will be quantified across representative ecosystems subjected to an array of land use activities. Site-specific information will be incorporated into hierarchical models for achieving global descriptions of the current conditions and for predicting the consequences of future changes in land use and climate. The general research issue will be addressed by the following four steps:

• Establish a network of accurate measurements of gaseous and hydrologic nutrient fluxes across landscapes representing major ecosystems that have received minimal impacts from human activities. These experimental landscapes will be paired with those receiving various land management practices (Gosz et al., 1988; Jordan et al., 1986; Lowrance et al., 1985; Peterjohn and Correll, 1984). Examples of land use changes that must be studied in appropriate regions include whole tree harvesting, introduction of multiple cropping and irrigation, conversions between forest-grassland-cropland, and various urbanization scenarios. Watersheds have proved to be powerful experimental approaches for biogeochemical studies (Likens et al., 1985) and should for a major component of these experiments. The watershed experimental designs permit an analysis of inputs and outputs to the research landscapes, and, in addition, the discharge streams become integrating measures of biogeochemistry of the area under study.

• Processes that control the fluxes of material will be described by ecosystem type. Models describing these processes will be driven by variables subject to climatic change, thus allowing a prediction of changes in flux rate as a function of global climatic change. To describe these fluxes and to predict both the effects of land use and global change on these fluxes and their feedback on global change, an understanding of various control factors must be included.

- Field tests and models will be developed relating fluxes of water, sediment, nutrients, and pollutants to interactive fluxes of trace gases between the biosphere and the atmosphere. Substantial progress has been made in the development of ecosystem models that simulate dynamics of material flows and nutrient cycling with plant production for specific systems (Parton et al., 1987; Pastor and Post, 1986). These models have been tested and validated using site-specific data. Recently, these models have been used to simulate regional ecosystem dynamics by linking ecosystem-level models to regional geographic information systems (GISs). The GIS systems contain information about the spatial variations in the soils, land use, and climatic variables for a region. The linked GIS and ecosystem model systems (Burke et al., 1990; Welch et al., 1988) use the GIS system to provide the driving variables for simulation of the dynamics of the plant and soil system for the spatial grid.

- Models will be developed for each region of the biosphere relating these material fluxes to land use and to climate variables. These models will then be used for identifying the regions and land uses most susceptible to climatic change.

Research Priorities. The following three priorities should be addressed by comprehensive studies.

- The fate and effects of nitrogen and other nutrients deposited on terrestrial ecosystems. Nitrogen is added to terrestrial systems by natural processes (e.g., lightning) and in large amounts through fertilizer application and as air pollutants. Once on the landscape, nitrogen accumulates in the vegetation and soil, and eventually in the groundwater, streams, estuaries, and oceans (Gildea et al., 1986; Kempe, 1988; Vorosmarty et al., 1986). Similar processes also add sulfur and other nutrients to terrestrial systems. The fate and effects of added nitrogen and other materials must be quantified in each of the components of major ecosystems, especially as they undergo changes in climate and land use. Particular attention must be paid to nitrogen because of its relationship with the global carbon budget, i.e., the temperate zone is both the recipient of significant amounts of nitrogen and a major sink for carbon dioxide. Thus these local processes are also important at regional and global scales.

- Nutrient transfers caused by cultivation. Under cultivation (e.g., cropping and forest harvest), nutrients are lost from terrestrial ecosystems. This loss of soil fertility has enormous impacts on the long-term productivity of the landscape and on the sustainable development of cropping systems.

Fertility loss is caused by the interactions of cropping systems, soil conditions, and the prevailing climate. Despite some understanding of nutrient loss under standard agricultural practices in some well-studied regions (Beaulac and Reckhow, 1982; Bowden and Bormann, 1986; Lowrance et al., 1985; Robertson and Tiedje, 1987; Schimel et al., 1985), these processes are not known for many major ecosystem types. These losses are to the atmosphere, via surface flow and into the groundwater. The research questions that must be answered are how are these materials distributed, what are the pathways of these materials, and what are the consequences to the recipient ecosystems, i.e., how are these chemicals processed in the recipient terrestrial and aquatic systems? Moreover, these distribution and processing questions must be answered for systems undergoing changes in land use and climate. These are regional processes that must be aggregated to the global scale because land use patterns are effective at the local scale, but climatic changes occur at broader scales.

- Aeolian and alluvial erosion. In arid and semiarid regions, and possibly in polar regions, there is a significant redistribution of earth surface materials by aeolian and alluvial processes (Schimel et al., 1985). Climatic change in the paleorecord has also been associated with these airborne and waterborne materials (see chapter 3). That is, dust in the record is an indicator of climatic change. In addition, changes in phosphorus, iron, and other materials by these erosional processes also affect freshwater, estuary, and ocean systems. Thus these aeolian and alluvial processes are both an indicator of climatic change and a consequence of changing climate. Research approaches include coupling processes occurring at various spatial scales. For example, rain simulators and shelters could be used to determine the effects of climatic change on erosional processes at the sources, and cesium-137 techniques may be applicable to determining the distribution patterns of the material that is moved by wind and water. It may be possible to link a particle sensing network for airborne materials to changes in characteristics of the ecosystem that can be monitored by remote sensing.

In all three of the research priorities, there is a need to link airsheds and watersheds. Landscapes are heterogeneous, and to answer these questions it is necessary not only to determine the net exchange of materials within and among ecosystems but also to understand how materials are processed in these heterogeneous regions. The study of large drainage basins, with heterogeneous land uses and natural features, will be an important research approach. These basins include biogeochemical processes in terrestrial ecosystems and in various impoundments, streams, and margins along different land use types. Materials are transferred, sequestered, and processed in different ways within the heterogeneous basins. The behavior of these basins and their interactions with the atmosphere must be understood on the global scale.

Fluxes from Terrestrial to Coastal Marine Systems

Estuaries and coastal seas are increasingly influenced by human activities and are rapidly being degraded in many regions of the world. Although toxic substances in estuaries are of major concern, much of the degradation of estuaries and coastal seas can be traced to land use practices, eutrophication, and the development of anoxic waters (Kemp et al., 1983, 1984; Larsson et al., 1985; Nixon, 1982; Officer et al., 1986; Price et al., 1985). Net primary production in many estuaries is limited by nitrogen, and so eutrophication of these systems is caused by excessive inputs of nitrogen (Boynton et al., 1982; D'Elia et al., 1986; Howarth, 1988). The export of nitrogen from terrestrial ecosystems ("nonpoint sources") accounts for half or more of the total nitrogen inputs to many major estuaries, such as Delaware Bay and Chesapeake Bay (Correll, 1987; Nixon and Pilson, 1983; Nixon et al., 1986). Thus any change in the functioning of terrestrial ecosystems over large scales is likely to have a major effect on the downstream estuarine and coastal marine ecosystems, in many ways the ultimate receivers of substances exported from these terrestrial ecosystems.

There are two major concerns of eutrophication in the coastal zone. These are the effect on the coastal production of fisheries and wildlife and the effect on atmospheric fluxes of trace gases, particularly nitrous oxide and dimethylsulfide. Hypoxic waters now occur on the Louisiana coastal shelf near the Mississippi Delta over areas as large as, or larger than, similar phenomena reported on the East Coast (Turner et al., 1987). The increase appears to be attributable to increased nitrogen delivery and reduced sediment delivery by the Mississippi River. Sediments in the Mississippi have been trapped by upstream dams, beginning in the 1950s (Meade and Parker, 1985). Decreased suspended sediments and increased water clarity and nitrogen are expected to increase phytoplankton production, which is light- and nitrogen-limited. Increased settling of phytoplankton, zooplankton, and organic material into bottom waters and removal of oxygen by decomposition may explain the expansion of the hypoxic zones (Turner et al., 1987). Virtually all large rivers of the world show similar patterns in sediment and nutrient delivery because of land use and dams. Resulting increases in phytoplankton production and settling in coastal zones may have a significant effect on the uptake of carbon dioxide from the atmosphere and long-term carbon storage. Coastal marine ecosystems have been postulated to be important sources of both nitrous oxide (Seitzinger et al., 1983) and dimethylsulfide (Andreae and Raemdonck, 1983). Fluxes of both of these can be expected to increase if rates of net primary production increase (Seitzinger et al., 1983; Andreae and Raemdonck, 1983).

The movement of nitrogen and other substances from terrestrial ecosystems to estuaries and coastal marine ecosystems may be altered by at least three factors: (1) changes in land use and stream flow by stream regulation,

(2) changes in atmospheric inputs to the terrestrial ecosystems, and (3) changes in climate. Each of these is briefly discussed below.

1. Land use. That changes in land use alter nutrient and sediment export from terrestrial ecosystems is well known in a qualitative sense, and yet this has only been well studied in a relatively few areas. Most studies have concentrated either on the effects of disturbance on element export from forested ecosystems (Bormann and Likens, 1979) or on export from agroecosystems (Beaulac and Reckhow, 1982; Lowrance et al., 1985). Some studies on element export from urban and suburban environments exist, but are of a site-specific nature. Some fairly sophisticated models are available for analyzing element export as a function of land use (Delwiche and Haith, 1983; Haith and Shoemaker, 1987), but these have been only partially tested, and their treatment of export from suburban and urban environments tends to be simplistic. One potential difficulty in applying such models to element export from terrestrial ecosystems to estuaries is that they do not allow for processing of substances within rivers, including riparian zones, natural main stem lakes, and man-made reservoirs, floodplains, and deltas (Costanza et al., 1990; Howarth et al., 1990; Mulholland, 1981; Vorosmarty et al., 1986).

2. Atmospheric inputs. A recent technical report from the Environmental Defense Fund (Fisher et al., 1988) concluded that nitrate in acid precipitation falling on terrestrial ecosystems can be a major source of nitrogen reaching estuaries. While this is a reasonable hypothesis, very little is known about the retention and export of nitrogen falling on terrestrial ecosystems in precipitation. Likens et al. (1985) found no clear relationship between nitrate inputs in precipitation and nitrate exports in stream flow over a period of 14 years in the Hubbard Brook watershed. Also, nitrogen exports from agricultural lands are not clearly related to nitrogen inputs in fertilizer (Beaulac and Reckhow, 1982), suggesting that export is also unlikely to be tightly related to input in precipitation. As discussed in the previous section, the factors regulating nitrogen retention and export from terrestrial ecosystems clearly deserve further study (Hooper et al., 1988).

3. Climate. It seems likely that the export of elements and sediment from terrestrial ecosystems to coastal marine ecosystems can be altered by changes in climate, but this has received little study. Increased erosion and increased runoff resulting from a wetter climate seem likely to greatly increase element export, although nitrogen export might be decreased if denitrification within soils increases markedly, a result of more waterlogged conditions. Results from a land-use, carbon- and sediment-export model for the Hudson River watershed suggest that carbon and sediment export may be more sensitive to the seasonal and day-to-day patterns in precipitation than to annual amounts (Howarth et al., 1989). This may be true for other elements. In addition to the direct effects of climatic change on element

export through changing erosion and runoff, climatic change is likely to alter the structure and function of terrestrial ecosystems, which also is likely to alter element exports.

Research Priorities. Three approaches to addressing issues related to substance movement from terrestrial ecosystems to coastal marine ecosystems are recommended:

• Establish a national surface water chemistry sampling network. The USGS operated a national surface water chemistry sampling network from the early 1970s until the early 1980s in conjunction with some of their gauging stations, but chemical sampling has largely been discontinued or has been extremely rare at most stations since the early 1980s. Such a network should be reestablished and expanded so as to better discern temporal variation in element export from terrestrial ecosystems to aquatic systems and so as to detect differences in export from different types of ecosystems and in different climatic regimes. The recently completed USGS 4-year pilot project to test concepts for a National Water Quality Assessment (NAWQA) program represented a diversity of hydrological environments and water quality conditions, and as such, would provide a useful initial model (Hirsch et al., 1988).

• Establish detailed watershed studies to examine element and substance export as a function of land use and climate. These studies would measure exports of carbon, nitrogen, phosphorus, water, and sediment at the watershed scale with a much greater sampling frequency than used in the national surface water chemistry network. Watersheds should be selected so as to represent different land uses (e.g., undisturbed forests, agriculture, and suburbia) in given climatic types. Within any given climatic types, the watersheds representing different land uses should be in close proximity and should have similar parent materials. These studies should be run long enough to determine the effects of year-to-year variability in climate on element export and resulting effects on aquatic systems. Studies should be established in three or four different climatic regimes.

• Develop improved models for the export of substances from terrestrial ecosystems and for movement of these substances to estuaries. Models for the movement of surface and subsurface water, sediment, and elements from various terrestrial ecosystems should be improved. Such models could be tested using data collected from the proposed national surface-water chemistry sampling network (first item above) and from the proposed watershed export studies (second item above). The goal of these models should be to allow better prediction of the potential effects of land use change, effects from alterations in atmospheric inputs to terrestrial ecosystems, and effects from climatic change on both terrestrial and aquatic systems.

METHODS AND INSTRUMENTS

Models

For most of the potentially important trace gases, there is still unacceptable uncertainty in the identity and global distribution of the main sources and sinks. Although the ultimate goal is to understand the fluxes of gases, nutrients, and pollutants, an essential first step is to quantify how concentrations of each important material vary geographically and temporally and then to relate these patterns to the distribution of ecosystem types and anthropogenic sources. This can be accomplished by a combination of monitoring to measure the material concentrations and the associated biotic and physical variables and modeling to develop the predictive multivariate relationships.

Models will be most effective when developed in parallel with standardized monitoring stations located in representative ecosystem types (including human-modified ecosystems containing anthropogenic sources) around the world. It would be most efficient if these stations measured the concentrations of all potentially important trace gases and other materials as well as the values of additional climatic and geological variables with sufficient accuracy and frequency to assess seasonal and interannual variation. Additional data on environmental variables, such as local vegetation types and land use categories, can be obtained from remote sensing and other sources (e.g., government census records). It is recommended that the modeling and monitoring be conducted so as to quantify spatial variation on the scale of patterns of vegetation and land use change across the landscape. A scale of approximately 1 km is recommended as a general guide, but the sample distribution pattern should be based on the statistical distributions of the putative driving variables.

It is anticipated that to understand global fluxes and their controls three scales of models (micro, meso, and macro) will have to be developed (see also chapters 2 and 5). In addition, because there will often be important spatial heterogeneity within each of these scales, it will be necessary to develop techniques to aggregate or synthesize the outputs of the models at smaller scales to use as inputs for models at larger scales.

- Microscale—Fluxes across the interface between the atmosphere and vegetation, soil, and water surfaces. Small-scale dynamics will vary with gas species or other materials and with environmental variables that affect sources, sinks, chemical reactions, and micrometeorological conditions. At this level it is possible to rely on basic knowledge of microbial physiology, transport processes, and chemistry. These models are often portable with minimal reparameterization but do require detailed input data. Small-scale experiments and models will be particularly important for elucidating the mechanisms that control production and deposition and for predicting the changes in controlling factors and the resulting fluxes that may accompany different types of global change.

For considerations of global changes in trace gases, it may not always be necessary to produce detailed, mechanistic models at this level. For example, for gases of primarily anthropogenic origin and readily identifiable sources, mesoscale models may suffice. But at least for some gas species with unknown biogenic sources and/or with moderate half-lives in the atmosphere, detailed studies of the near-surface dynamics will be essential. A likely example is carbon monoxide, which appears to be produced in significant quantities in at least some tropical forests and to exhibit vertical changes in concentration between the soil and the canopy.

• Mesoscale—Fluxes within patches of similar ecosystem type. At the scale of approximately 1 km, it should be relatively easy to use remotely sensed and ground-based data to classify ecosystem types, including heavily human-modified ones such as different kinds of agricultural, suburban, and urban systems. It should also be practical, for example, to monitor spatial and temporal variation in gas concentrations at this scale (i.e., in the lower atmosphere above the vegetation). What is needed are predictive process models to characterize the sources, sinks, chemical transformations, and fluxes of gases and other materials that occur within three-dimensional cells at this scale. Moreover, we need to know, for example, how much biological detail is needed to characterize fluxes from physiognomic types of vegetation.

Figure 6.3 illustrates the five essential components of a mesoscale model: (1) vertical exchange with the soil, water, or vegetation that covers the earth's surface (inputs characterizing these production and deposition processes come from the microscale models, appropriately aggregated if necessary to account for surface heterogeneity); (2) vertical exchange with the upper atmosphere; (3) horizontal exchange, via wind, with adjacent patches of the same or different ecosystem type; (4) circulation and chemical reactions within the cell that affect the concentration and flux; and (5) environmental forcing functions, such as changes in vegetation, temperature, cloud cover, or the concentrations of other materials, that are likely to change the dynamics of the gas or nutrient species in question.

Models must be customized to account for the unique features of each gas nutrient or pollutant species. The local production and deposition components and the circulation and chemical reactions within the cell will tend to be species specific.

• Mesoscale and macroscale—Linking trace gas process models to earth system models. The final state in modeling global fluxes is to aggregate the mesoscale cells and incorporate the trace gases and other materials into atmosphere-biosphere interface models. This is necessary not only to account

190 RESEARCH STRATEGIES FOR THE USGCRP

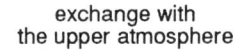

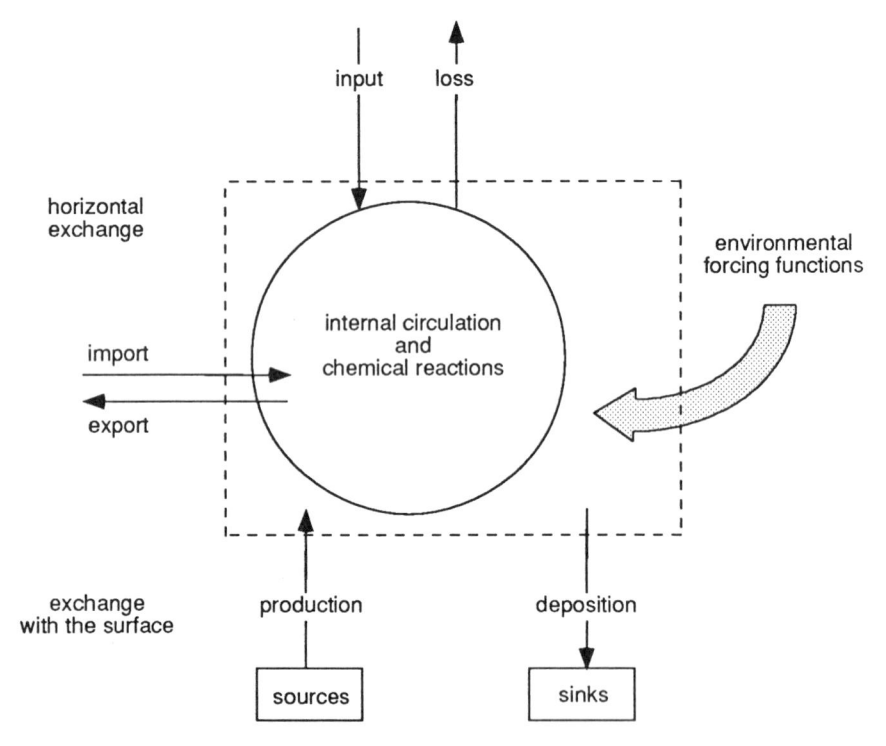

FIGURE 6.3 Main ingredients of a mesoscale model for trace gas fluxes.

for regional variations in concentration and long-distance transport of anthropogenic gases and other materials, but also to understand the global fluxes of any species for which the primary sources and sinks may be spatially isolated (e.g., between tropical forests and oceans).

The problems of predicting atmospheric circulation on a scale of approximately 100 km have not been solved, but there is currently a major research effort to improve and test GCMs. These models can be used (Matthews and Fung, 1987) to predict the large-scale dispersal of trace gases if atmospheric reactions are included. A much more difficult problem would seem to be the development of techniques for aggregating the outputs of mesoscale models over heterogeneous landscapes (see chapter 5) to obtain accurate inputs for the GCMs. For example, construction of an atmospheric boundary layer model should include (1) drag coefficients that vary with topography and vegetation especially when topographic variation is relatively small, (2) turbulent exchange coefficients that depend on these drag coefficients and on the thermal structure of the lower atmosphere, and (3) moisture-balance equations that depend on current and changing vegetation characteristics (Walker et al., 1990).

High priority should be placed on developing mesoscale models of trace gas flux, and then aggregating or synthesizing them over space so that they can be used as inputs into GCMs. More work on this type of predictive modeling is required, and for the full suite of gases and materials of interest. To avoid doing such modeling in an information vacuum, it will be necessary to accompany this effort with systematic monitoring of spatial and temporal variation in trace gas concentrations and material fluxes as a function of ecosystem type and with initial descriptive models that quantify the environmental correlates of this variation.

Lower priority should be placed on developing microscale and macroscale models, because there is already considerable effort at these levels. But the pace of research at micro- and macro-levels must also be increased if we are to produce predictive models of global trace gas fluxes in time to deal with these and other pressing problems of global change.

Instrumentation for Measuring Fluxes

One of the current key limitations in formulating a predictive understanding of global processes is the inability to measure unequivocally the abundances of many trace species that are centrally involved in those processes. This is particularly true for the measurement of chemical fluxes (emission or deposition), where the number of gas species for which it is generally accepted that reliable techniques exist are only a few.

A wide variety of emission sources, deposition surfaces, and chemical species are involved in global fluxes. Natural emissions of chemically or climatically important compounds occur from terrestrial and oceanic sources (e.g., nonmethane hydrocarbons and methane from wetlands). Deposition surfaces that figure strongly in major removable processes range from vegetative uptake (e.g., of carbon dioxide) to physical attachment (e.g., nitric acid depositing on wetted soils). The chemical variety of the emitted or depositing compounds (inert species and reactive radicals) implies that the likelihood of even semiuniversal detectors is unlikely.

Flux measurements of trace gases (molecules per unit area per unit time) require a determination of the atmospheric concentrations over time. Measurement of low concentrations of many chemical compounds requires highly sensitive detectors and rigorous analytical quality assurance. The need to obtain a representative flux from a large spatial area generally implies use of remote or aircraft sensing. Many natural emissions are quite sensitive to moisture, temperature, and other such factors, thereby introducing substantial spatial and temporal variability not well studied in most contemporary investigations. Because of the challenges that flux measurements pose and because of the necessity to substantially improve the current capabilities, a focused program for new methodologies and instrumentation is needed.

To date, flux measurements have been made in enclosures, along gradients, and via eddy correlation. The enclosure method establishes the flux from a small area based on increases in concentration of the compound in the container. The gradient method generally employs towers to determine differences in the target compound or element across a spatial gradient (e.g., as a function of altitude, in conjunction with meteorological analyses). The eddy correlation method, often used with towers or aircraft, relates small-scale concentration variations to variations in air motion. The usefulness of these methods depends on the research question and the scale of the investigation.

The key to success in the gradient and eddy correlation methods is the availability of rapid-response sensors. Detectors are needed that can make reliable measurements of the concentration of a species at the part-per-trillion level and with a measurement rate of less than once per second. Thus the development of new physical and chemical sensors with those characteristics is the key to improving the status of flux measurements. Improvements and new innovations in instruments that measure more than one species simultaneously are especially needed, since covariation provides key insight into the biogenic processes involved.

It is imperative that rigorous intercomparison experiments precede the widespread and large-scale application of flux measurement techniques. The atmospheric chemistry community has developed an approach that provides a valuable unbiased indication of measurement capabilities. The main features of the experiments that have proved the most informative are the following:

- several different methods for measuring the same species are involved;
- "mature" instruments (i.e., those that have been used in published investigations) are compared;
- measurements are made at the same time and place and under typical and documented environmental conditions, insofar as possible;
- the expected accuracy and precision are hypothesized in advance of the intercomparison; and
- all results and conclusions are published in the open literature.

CROSS-CUTTING ISSUES

Trace gas sources and sinks are affected by the intrinsic characteristics of ecosystems, by changes in land use, and by changing climatic conditions.

Understanding both the reservoirs and the fluxes under these three conditions is necessary for documenting and predicting global change. These studies require careful stratification to ensure a parsimonious set of measured conditions with the greatest experimental efficiency. Thus trace gas studies will profit from a comprehensive experimental design that addresses several gases simultaneously. Similarly, in many cases nutrient flux studies can be conducted with trace gas studies. Finally, as discussed in chapter 5, water-energy balance of the biosphere will require instrumented watersheds. These, too, can be combined with the trace gas and nutrient flux studies.

A global data base of direct measurements of trace gas fluxes to and from ecosystems is not achievable in the foreseeable future. No technology is available that would allow such measurements to be made remotely from satellites, and global surveys using ground-based or aircraft platforms would involve tremendous costs and logistical difficulties. The best approach at this time for constructing a global data base of trace gas fluxes is to map the environmental variables known to regulate those fluxes from each ecosystem. The functional dependences relating trace gas fluxes to these variables can then be quantified by ground-based and aircraft studies focusing on specific ecosystems, and the resulting data can then be aggregated as appropriate. Laboratory and small-scale experiments must be conducted for the purpose of relating trace gas fluxes to input variables that can be measured via remote sensing techniques.

Since measuring flux rates directly is difficult at global scales and grids of concentration data are much more feasible, it will be necessary to develop mathematical methods for inverting from concentration data to flux rates, and to be able to do so at local to regional to global scales. Additional constraints on the inversions may be derived from remote sensing and the development of large-scale soil and land use data bases. Also, there is a need to refine statistical techniques that identify adequate sample sizes in relation to the cost of acquiring data and the required sampling intensity.

Large manipulation experiments will be necessary under selected conditions that represent important types of ecosystems, e.g., agricultural systems. In other instances, where the initial conditions are variable and heterogeneous, comparative measurements may be more reasonable. Moreover, careful analysis will be required of existing data, both to synthesize what is already known and for designing efficient experiments. Connecting large-scale manipulation experiments with experimental Regional Research Centers will contribute to research economy and assist in the extrapolation of results to regional and global scales.

Nutrient transfer studies measuring the lateral fluxes of nutrients should be organized to include hierarchical descriptors of land use arrangements and other driving variables. Most of the experimental conditions are in place, and thus the challenge is to establish the field measurements and not to develop new large-scale experimental conditions. The most difficult step in some of the studies will involve the prediction of how these fluxes will change with alterations in the regional and global climate. Thus the scenarios for global climatic change must be solidified as a basis for these experiments and for subsequent models.

Great economies can be achieved by careful coordination of the nutrient transfer and trace gas flux measurements. Making the measurements at the same sites will minimize logistic expenses and assist in the development of microscale and mesoscale models and of correlative indicators of system dynamics and responses.

In many instances, the first step in these studies will be to develop initial models to determine unknown parameters and to identify the experiments most likely to yield critical important information. Thus models will be important at all stages of the studies, from beginning synthesis of known information and experimental design to final synthesis of new information and for scaling among time and space scales.

These studies on trace gases and the fluxes of materials involve a wide spectrum of traditional disciplines and will require a significant number of scientists over one to two decades. In addition, the investigations will depend on a thorough understanding of human systems, especially in terms of land use practices. Thus the scientific community must be sure that there are educational programs that include this wide spectrum of disciplines. There must also be undergraduate and graduate programs that encourage the best of our students to participate in these studies.

REFERENCES

Andreae, M.O. 1985. The emission of sulfur to the remote atmosphere: Background paper. Pp. 5-26 in J.N. Galloway et al. (eds.), The Biogeochemical Cycling of Sulfur and Nitrogen in the Remote Atmosphere. D. Reidel, Dordrecht, The Netherlands.

Andreae, M.O., and H. Raemdonck. 1983. Dimethyl sulfide in the surface ocean and the marine atmosphere: A global view. Science 221:744-747.

Andreae, M.O., et al. 1988. Biomass burning emission and associated haze layers over Amazonia. J. Geophys. Res. 93:1509-1527.

Andreae, M.O., H. Berresheim, H. Bingemer, D.J. Jacob, and R.W. Talbot. 1990. The atmospheric sulfur cycle over the Amazon Basin. 2. Wet season. J. Geophys. Res., in press.

Beaulac, M.N., and K.H. Reckhow. 1982. An examination of land use-nutrient export relationships. Water Res. Bull. 18:1013-1024.

Billings, W.D. 1987. Carbon balance of Alaskan tundra and taiga ecosystems: Past, present and future. Quaternary Science Reviews 6:1265-1277.

Bormann, F.H., and G.E. Likens. 1979. Pattern and Process in a Forested Ecosystem. Springer-Verlag, New York.

Bowden, W.B., and F.H. Bormann. 1986. Transport and loss of nitrous oxide in soil water after forest clearcutting. Science 223:867-869.
Boynton, W.R., W.M. Kemp, and C.W. Keefe. 1982. A comparative analysis of nutrients and other factors influencing estuarine phytoplankton production. Pp. 69-90 in V.S. Kennedy (ed.), Estuarine Comparisons. Academic Press, New York.
Burke, I.C., D.S. Schimel, C.M. Yonker, W.J. Parton, and L.A. Joyce. 1990. Regional modeling of grassland biogeochemistry using GIS. Landscape Ecology 4, in press.
Committee on Earth Sciences (CES). 1989. Our Changing Planet: The FY 1991 Research Plan. The U.S. Global Change Research Program. Washington, D.C.
Conrad, R., and W. Seiler. 1985. Influence of temperature, moisture, and organic carbon on the flux of H_2 and CO between soil and atmosphere: Field studies in subtropical regions. J. Geophys. Res. 90:5699-5709.
Correll, D.E. 1983. N and P in soils and runoff in three coastal plain land uses. Pp. 207-224 in R.R. Lowrance, R.L. Todd, L. Asmussen, and R.A. Leonard (eds.), Nutrient Cycling in Agricultural Ecosystems. Special Publication No. 23. Agriculture Experiment Station, University of Georgia, Athens, Ga.
Correll, D.L. 1987. Nutrients in Chesapeake Bay. Pp. 298-302 in S.K. Majumbar, L.W. Hall, Jr., and H.M. Austin (eds.), Contaminant Problems and Management of Living Chesapeake Bay Resources. Pennsylvania Academy of Sciences, Gettysburg, Pa.
Costanza, R., F.H. Sklar, and M.L. White. 1990. Modeling coastal landscape dynamics. BioScience 40:91-107.
D'Elia, C.F., J.G. Sanders, and W.R. Boynton. 1986. Nutrient enrichment studies in a coastal plain estuary: Phytoplankton growth in large-scale, continuous cultures. Can. J. Fish. Aquat. Sci. 43:397-406.
Delmas, R., and J. Servant. 1983. Atmospheric balance of sulfur above an equatorial forest. Tellus 35:110-120.
Delwiche, L.L.D., and D.A. Haith. 1983. Loading functions for predicting nutrient losses from complex watersheds. Water Res. Bull. 19:951-959.
Fisher, D., J. Ceraso, T. Mathew, and M. Oppenheimer. 1988. Polluted Coastal Waters: The Role of Acid Rain. Environmental Defense Fund, New York.
Galbally, I. (ed.). 1989. The International Global Atmospheric Chemistry (IGAC) Program: A Core Project of the International Geosphere-Biosphere Program. IAMAP Commission on Atmospheric Chemistry and Global Pollution. Renwick Pride Pty Ltd., Albury, Australia. 55 pp.
Gildea, M.P., B. Moore, C.J. Vorosmarty, B. Bergquist, J.M. Melillo, K. Nadelhoffer, and B.J. Peterson. 1986. A global model of nutrient cycling: I. Introduction, model structure and terrestrial mobilization of nutrients. Pp. 1-31 in D.L. Correll (ed.), Watershed Research Perspectives. Smithsonian Press, Washington, D.C.
Goldan, P.D., R. Fall, W.C. Kuster, and F.C. Fehsenfeld. 1988. Uptake of COS by growing vegetation. J. Geophys. Res. 93:186-192.
Gosz, J.R., C.N. Dahm, and P.G. Risser. 1988. Long-path FTIR measurement of atmospheric trace gas concentrations. Ecology 69:1326-1330.

Haith, D.A., and L.L. Shoemaker. 1987. Generalized watershed loading functions for stream-flow nutrients. Water Res. Bull. 23:471-478.

Hanst, P.L., J.W. Spence, and E.O. Edney. 1980. Carbon monoxide production in photooxidation of organic molecules in the air. Atmos. Environ. 14:1077-1088.

Hirsch, R.M., W.M. Alley, and W.G. Wilber. 1988. Concepts for a national water-quality assessment program. U.S. Geological Survey Circular 1021. 42 pp.

Hooper, R.P., A. Stone, N. Christopherson, E. de Grosbois, and H.M. Seip. 1988. Assessing the Birkenes model of stream acidification using a multi-signal calibration methodology. Water Resour. Res. 24:1308-1316.

Howarth, R.W. 1988. Nutrient limitation of net primary production in marine ecosystems. Annu. Rev. Ecol. Syst. 19:89-110.

Howarth, R.W., J.R. Fruci, and D. Sherman. 1990. The influence of land use on functioning of estuarine ecosystems: The Hudson River estuary as a case study. Ecological Applications, in press.

Jordan, T.E., D.L. Correll, W.T. Peterjohn, and D.E. Weller. 1986. Nutrient flux in a landscape: The Rhode River watershed and receiving waters. Pp. 57-76 in D.L. Correll (ed.), Watershed Research Perspectives. Smithsonian Press, Washington, D.C.

Kemp, W.M., W.R. Boynton, R.R. Twilley, J.C. Stevenson, and J.C. Means. 1983. The decline of submerged vascular plants in upper Chesapeake Bay: Summary of results concerning possible causes. Mar. Technol. Soc. J. 17:78-89.

Kemp, W.M., W.R. Boynton, R.R. Twilley, J.C. Stevenson, and L.G. Ward. 1984. Influences of submerged vascular plants on ecological processes in upper Chesapeake Bay. Pp. 367-394 in V.S. Kennedy (ed.), The Estuary as a Filter. Academic Press, New York.

Kempe, S. 1988. Estuaries—their natural and anthropogenic changes. Pp. 251-285 in T. Rosswall, R.G. Woodmansee, and P.G. Risser (eds.), SCOPE 35: Scales and Global Change. John Wiley and Sons, Chichester, England.

Larsson, U.R., R. Elmgren, and F. Wulff. 1985. Eutrophication and the Baltic Sea: Causes and consequences. Ambio 14:10-14.

Lenschow, D.H., and B.B. Hicks (eds.). 1989. Global Tropospheric Chemistry: Chemical Fluxes in the Global Atmosphere. Report of the Workshop on Measurements of Surface Exchange and Flux Divergence of Chemical Species in the Global Atmosphere. National Center for Atmospheric Research, Boulder, Colo.

Likens, G.E., F.H. Bormann, R.S. Pierce, and J.S. Eaton. 1985. The Hubbard Brook Valley. Pp. 9-39 in G.E. Likens (ed.), An Ecosystem Approach to Aquatic Ecology. Springer-Verlag, New York.

Logan, J.A. 1985. Tropospheric ozone: Seasonal behavior, trends and anthropogenic influence. J. Geophys. Res. 90:10463-10482.

Logan, J.A., M.J. Prather, S.C. Wofsy, and M.B. McElroy. 1981. Tropospheric chemistry: A global perspective. J. Geophys. Res. 86:7210-7254.

Lowrance, R.R., R.A. Leonardd, and L.E. Asmussen. 1985. Nutrient budgets for agricultural watersheds in the southeastern coastal plain. Ecology 66:287-296.

Marenco, A., and J.C. Delaunay. 1985. Experimental evidence of natural sources of CO from measurements in the troposphere. J. Geophys. Res. 85:5599-5613.

Matthews, E., and I. Fung. 1987. Methane emission from natural wetlands: Global distribution, area, and environmental characteristics of sources. Global Biogeochemical Cycles 1:61-86.

Meade, R.H., and R.S. Parker. 1985. Sediment in rivers of the United States. Pp. 49-69 in National Water Summary 1984. U.S. Geological Survey Water Supply Paper 2275.

Meyers, T.P., and D.D. Baldocchi. 1988. A comparison of models for deriving dry deposition fluxes of O_3 and SO_2 to a forest canopy. Tellus 40:270-284.

Mooney, H.A., P.M. Vitousek, and P.A. Matson. 1987. Exchange of materials between terrestrial ecosystems and the atmosphere. Science 238:926-932.

Mulholland, P.J. 1981. Deposition of riverborne organic carbon in floodplain wetlands and deltas. Pp. 142-172 in Carbon Dioxide Effects Research and Assessment Program: Flux of Organic Carbon by Rivers to the Oceans. Report of a Workshop held in Woods Hole, Mass., September 21-25, 1980. Committee on Flux of Organic Carbon to the Ocean, Division of Biological Sciences, National Research Council. U.S. Department of Energy, Office of Energy Research, Washington, D.C. 397 pp.

National Research Council. 1984. Global Tropospheric Chemistry: A Plan for Action. National Academy Press, Washington, D.C.

National Research Council. 1988. Toward an Understanding of Global Change: Initial Priorities for U.S. Contributions to the International Geosphere-Biosphere Program. National Academy Press, Washington, D.C. 213 pp.

Nixon, S.W. 1982. Nutrient dynamics, primary production, and fisheries yields of lagoons. Pp. 357-371 in Oceanologica Acta. Special publication, Proceedings of the International Symposium on Coastal Lagoons, SCOR/IABO/UNESCO, 1981, Bordeaux, France.

Nixon, S.W., and M.E.Q. Pilson. 1983. Nitrogen in estuarine and coastal marine ecosystems. Pp. 565-648 in E.J. Carpenter and D.G. Capone (eds.), Nitrogen in the Marine Environment. Academic Press, New York.

Nixon, S.W., C. Oviatt, J. Frithsen, and B. Sullivan. 1986. Nutrients and the productivity of estuarine and coastal marine ecosystems. J. Limnol. Soc. South Africa 12:43-71.

Officer, C.B., R.B. Biggs, J. Taft, L.E. Cronin, M.A. Tyler, and W.R. Boynton. 1986. Chesapeake Bay anoxia: Origin, development, and significance. Science 232:22-27.

Parton, W.J., D.S. Schimel, C.V. Cole, and D. Ojima. 1987. Analysis of factors controlling soil organic levels in grasslands in the Great Plains. Soil Sci. Soc. Am. J. 51:1173-1179.

Pastor, J., and W.M. Post. 1986. Influence of climate, soil moisture and succession on forest carbon and nitrogen cycles. Biogeochemistry 2:3-27.

Peterjohn, W.T., and D.L. Correll. 1984. Nutrient dynamics in an agricultural watershed: Observations of the role of a riparian forest. Ecology 65:1466-1475.

Price, K.S., D.A. Flemer, J.L. Taft, and G.B. Mackiernan. 1985. Nutrient enrichment of Chesapeake Bay and its impact on the habitat of striped bass: A speculative hypothesis. Trans. Am. Fish. Soc. 114:97-106.

Robertson, G.P., and J.M. Tiedje. 1987. Deforestation alters denitrification in a lowland tropical rain forest. Nature 336:441-445.

Schimel, D.S., M.A. Stillwell, and R.G. Woodmansee. 1985. Biogeochemistry of C, N and P in a soil catena of the shortgrass steppe. Ecology 66:276-282.

Schimel, D.S., W.J. Parton, F.J. Adamsen, R.G. Woodmansee, R.L. Senft, and M.A. Stillwell. 1986. The role of cattle in the volatile loss of nitrogen from a shortgrass steppe. Biogeochemistry 2:39-52.

Schimel, D.S., M.O. Andreae, D. Fowler, I.E. Galbally, R.C. Harriss, D. Ojima, H. Rodhe, T. Rosswall, B.H. Svensson, and G.A. Zavarzin. 1989. Research priorities for studies on trace gas exchange. Pp. 321-331 in M.O. Andreae and D.S. Schimel (eds.), Exchange of Trace Gas Between Terrestrial Ecosystems and the Atmosphere. John Wiley and Sons, Chichester, England.

Seiler, W. 1978. The influence of the biosphere on the atmospheric CO and H_2 cycles. Pp. 773-810 in W.E. Krumbein (ed.), Environmental Biogeochemistry and Geomicrobiology, Vol. 3, Methods, Metals, and Assessment. Ann Arbor Science Publishers, Ann Arbor, Mich.

Seitzinger, S.P., M.E.Q. Pilson, and S.W. Nixon. 1983. Nitrous oxide production in nearshore marine sediments. Science 222:1244-1246.

Strain, B.R., and J.D. Cure. 1985. Direct effects of increasing carbon dioxide on vegetation. Duke University Press, Durham, N.C.

Tingey, D.T., M. Manning, L.C. Grothaus, and W.F. Burns. 1979. The influence of light and temperature on isoprene emission rates from live oak. Physiol. Plant. 47:112-118.

Tissue, D., and W.C. Oechel. 1987. Responses of *Eriophorum vaginatum* to elevated CO_2 and temperature in the Alaskan tussock tundra. Ecology 68:401-410.

Trabalka, J.R., and D.E. Reichle (eds.). 1986. The Changing Carbon Cycle: A Global Analysis. Springer-Verlag, New York.

Turner, R.E., R. Kaswadji, N.N. Rabalais, and D.F. Boesch. 1987. Long-term changes in the Mississippi River water quality and its relationship to hypoxic continental shelf waters. Pp. 261-266 in M.P. Lynch and K.L. McDonald (eds.), Estuarine and Coastal Management. Tools of the Trade. Proceedings of the Tenth National Conference of the Coastal Society, October 12-15, 1986, New Orleans, La. 391 pp.

Vorosmarty, C.J., M.P. Gildea, B. Moore, B.J. Peterson, B. Bergquist, and J.M. Melillo. 1986. A global model of nutrient cycling: II. Aquatic processing, retention and distribution of nutrients in large drainage basins. Pp. 32-56 in D.L. Correll (ed.), Watershed Research Perspectives. Smithsonian Environmental Research Center, Edgewater, Md. 421 pp.

Walker, B.H., S.J. Turner, R.J. Prinsley, D.M. Stafferd Smith, and H.A. Nix (eds.). 1990. Proceedings of the Workshops of the Coordinating Panel on Effects of Global Change on Terrestrial Ecosystems. I. A Framework for Modeling the Effects of Climate and Atmospheric Change on Terrestrial Ecosystems, Woods Hole, Mass., April 15-17, 1989. II. Non-modeling Research Requirements for Understanding, Predicting, and Monitoring Global Change, Canberra, August 29-31, 1989. III. The Impact of Global Change on Agriculture and Forestry, Yaounde, November-December 1989. IGBP Report No. 11. Stockholm, Sweden.

Welch, R., M. Remillard, and R.B. Slack. 1988. Remote sensing and GIS techniques for aquatic resource evaluation. Photogramm. Eng. Remote Sensing 54:177-185.

Williams, W.E., K. Garbutt, F.A. Bazzaz, and P.M. Vitousek. 1986. The response of plants to elevated CO_2. IV. Two deciduous-forest tree communities. Oecologia 69:454-459.

Yokouchi, Y., and Y. Ambe. 1984. Factors affecting the emission of monoterpenes from red pine (*Pinus densiflora*). Plant. Physiol. 75:1009-1012.

7
Biogeochemical Dynamics in the Ocean

OVERVIEW

In its 1988 report, *Toward an Understanding of Global Change: Initial Priorities for U.S. Contributions to the IGBP*, the Committee on Global Change recommended a research initiative on oceanic biogeochemical cycles (NRC, 1988). The objective of the effort was to develop the capability to predict the effect of projected climatic change on the ocean's physical, chemical, and biogeochemical processes, especially as they feed back to climate via the release or absorption of radiatively important gases such as carbon dioxide and organic sulfur species. This chapter identifies the current efforts to meet that challenge, both with existing programs and with recommendations for new efforts.

Global change is not limited to the physical aspects of climate. It affects, and is affected by, living processes. Today we know that the earth works as a system, and that physical, chemical, biological, and geological processes all interact to yield the constantly changing system of our environment. We also know that the ocean plays a key role in these interactive processes. For example, the importance of the ocean in the biogeochemical

---

This chapter was co-authored for the Committee on Global Change by D. James Baker, Joint Oceanographic Institutions, Inc., Chair; P. Brewer, Woods Hole Oceanographic Institution; H. Ducklow, University of Maryland; J. McCarthy, Harvard University; M. Reeve, National Science Foundation; and B. Rothschild, University of Maryland, with further comments by N. Andersen, National Science Foundation; K. Bryan, Princeton University; A. Jochens, Texas A&M University; W. Nowlin, Texas A&M University; J. O'Brien, Florida State University; J. Price, Woods Hole Oceanographic Institution; and C. Wunsch, Massachusetts Institute of Technology.

cycles of all of the elements essential to life on earth has been recognized for a long time. But the mechanisms that cause these cycles and their interaction with environmental change are not well understood. Moreover, the nature and even the sign of the possible feedbacks between environmental change and biogeochemical cycles driven by the activities of living organisms are unknown.

An example that shows how biological processes in the ocean can affect global change comes from the aerosols that are formed in the atmosphere over the ocean from the oxidation products of dimethylsulfide, a gas emitted by marine phytoplankton. Many of the physical properties of low-level clouds are dependent on the properties and distribution of the aerosols upon which the cloud droplets are formed. The aerosols affect the reflectivity, lifetime, and precipitation properties of these clouds. Because climatic factors may affect the activity of the marine phytoplankton, there is the possibility of a climatic feedback loop through the formation of these aerosols. In order to understand this process, efforts must be made to incorporate microphysical and chemical influences on cloud processes in global models. The relationships between phytoplankton activity, the emission of dimethylsulfide to seawater and subsequently to the atmosphere, the oxidation of dimethylsulfide to produce sulfate aerosols, and the relation of these aerosols to cloud condensation nuclei and the albedo of the earth need to be clarified, including the overall relationship to climatic changes (Charlson et al., 1987).

This is just one example of a specific linkage between oceanic chemistry, biology, atmospheric chemistry, and climate that underscores why, if we are to understand the cycles of the chemical elements, we must first understand their uptake and reactions with the ocean and its ecosystems. Another comes from the importance of the trace elements such as iron in determining global rates of phytoplankton new production. Offshore Pacific water, for example, appears to require supplemental iron from the atmosphere or continental margin to be optimally suited for plankton growth (Martin et al., 1990). Yet another example can be taken from the effect of ozone depletion on antarctic organisms. The potential increased levels of ultraviolet radiation could affect the phytoplankton that constitute the base of the food web in aquatic ecosystems by reducing the amount of primary production and altering community structure.

A principal practical concern is the role that the interactions of atmospheric and oceanic physical and chemical dynamics have on the long-term fluctuations of animal populations in the oceans, coastal seas, and estuaries. Whole economies are dependent on and sensitive to the interannual and decadal regional fluctuations of harvestable biomass, such as those correlated with El Niño. We need to develop an understanding of the complex interplay of biological and physical forcing on life history stages of animal populations, which can produce very large swings in the harvest of major components of the global ocean biomass from decade to decade.

This chapter focuses on five areas of biogeochemical dynamics in the ocean: biogeochemical fluxes, with an emphasis on carbon; the ocean-atmosphere interface; the oceanic ecosystem response to climatic change; the underlying physical processes in the ocean and atmosphere; and processes in the polar regions. The status of ongoing and proposed programs and ways in which they can be enhanced are discussed. Studies of biogeochemical processes in the coastal regions and air-sea fluxes, in particular, need further work. The upper ocean also needs more study, both in the area of physics and in biological processes.

Support for long-term monitoring in situ and by satellite-borne instruments, with an emphasis on carbon dioxide, ocean color, circulation, and winds, is strongly urged. It will also be important to provide support for the near-term research satellite missions in the early 1990s and the Earth Observing System (EOS) in the late 1990s to obtain the necessary data. The importance of long-term monitoring for physical, biological, and chemical variables in the ocean is underscored. The history of physical and biological events such as El Niño needs to be extended as far back in history as possible, using the proxy record as well as documentation, in order to define the statistical variability of these events (see chapter 3). Finally, the committee urges the development of improved ties among programs.

STATUS OF EXISTING EFFORTS

Biogeochemical Fluxes

The need for understanding biogeochemical cycles in the ocean led to a number of focused studies and advances in measurement capability in the 1970s and 1980s. The capability for global measurement of ocean color by satellite-borne instruments and the subsequent inference of biological productivity have provided new impetus to these studies. Also developed were in situ techniques for direct measurement of the vertical transport of biogenic material in the water column by sediment traps as well as high-precision methods for the detection of trace species in very small amounts.

The understanding and new ideas from the various programs and the capabilities provided by the new techniques led to the development of the international Joint Global Ocean Flux Study (JGOFS) (SCOR, 1987). The U.S. program has its counterparts in several other nations, including the U.K., France, Germany, and Japan. The goal of JGOFS is to determine and understand on a global scale the processes controlling the time-varying fluxes of carbon and associated biogenic elements in the ocean and to evaluate the related exchanges with the atmosphere, sea floor, and continental boundaries. JGOFS has been identified as a core project of the IGBP.

The JGOFS organizers have identified the special importance of carbon dioxide studies and the need for satellite observations of ocean color for determining the amounts of biological production in the ocean. To carry out a global survey of carbon dioxide in seawater, measurements of dissolved carbon dioxide will be made by JGOFS scientists during the global survey of the World Ocean Circulation Experiment (WOCE; see the section "Global Ocean Circulation" below). A series of intensive regional process studies in different ocean basins is also planned. The new understanding of biogeochemical mechanisms gleaned in these studies will be used to construct models for linking remotely sensed data from the surface of the ocean to large-scale views of elemental cycling in the oceanic interior. A JGOFS pilot study was conducted in the North Atlantic in the spring of 1989, and a major Pacific study is being planned.

The ocean color observation study, one of the major parts of the JGOFS program, is now in jeopardy. Detailed planning has taken place for the design construction, and flight in 1991 or 1992 of an ocean color sensor, the Sea-Viewing Wide Field Sensor (SEAWIFS). There is no other way to achieve the required global measurements of primary productivity on the appropriate spatial and temporal scales. Unfortunately, the plans for flight of this new ocean color instrument have been stalled because of a lack of commitment and funds. Unless a commitment can be made, the JGOFS field programs will lack the overall global satellite coverage that is vitally needed for understanding the biological productivity of the ocean and its effect on the interaction of biogeochemical processes with global climatic change. The present status is that there will be no ocean color measurements on a global scale until those from the ocean color instrument on the Japanese Advanced Earth Observation Satellite in 1995 or 1996. From 1997 on, ocean color will be monitored by the special high-spectral-resolution instruments on the EOS polar platforms. But these measurements will be too late to support the JGOFS field programs scheduled for 1990 to 1995.

JGOFS was originally planned to study all the various boundaries of the ocean, but it has had to focus on fewer of these. It is addressing the ocean-sediment interface, but it does not address in detail either the coastal interface or the ocean-atmosphere interface. As a consequence, JGOFS, now primarily aimed at the open ocean, will not address all of the issues related to biogeochemical dynamics in the ocean. Of critical interest are the coastal regions. It is in these regions that much of the biological productivity takes place, and yet it has been difficult to define a program because of the complexity of processes there. The committee believes that coastal regions should be a focus for the next phase of planning. Nationally, such planning could be a part of the Coastal Physical Oceanography (CoPO) program as it becomes interdisciplinary. Such planning for studies of interdisciplinary coastal processes is now a part of international planning with both the IGBP and the Intergovernmental Oceanographic Commission (IOC). The IGBP is now planning a Coastal Ocean Flux and Resource Study to help address these issues.

It should also be noted that the Department of Energy is reshaping its entire program dealing with coastal areas to specifically address interdisciplinary issues. A number of European efforts also address the coastal areas and are complementary to JGOFS. In addition, NOAA has developed a coastal program, and the EPA recently held a workshop on coastal ocean physics and climate change, aimed at the issues involved in the assessment of ecosystem response in the coastal ocean.

Ocean-Atmosphere Interface

Fluxes across the sea surface must also be studied. The atmosphere can provide an important transfer path for natural and pollution-derived chemicals entering the ocean. Some atmospherically derived species, especially nutrients like iron, may have important impacts on productivity in some areas of the ocean. The ocean is also an important source for several chemicals in the atmosphere, including sulfur species, certain low molecular hydrocarbons, and some halogenated species. It is important that the JGOFS program work closely with atmospheric chemistry programs so that fluxes from the atmosphere are measured and the mechanisms for transfer understood. Mechanisms to encourage such close interaction between programs need to be established. The physical fluxes of heat, water, and momentum are also important; these are discussed in the section "Physical Processes" (below).

During the late 1970s, there was a recognition by atmospheric chemists that trace species in the atmosphere including methane, nitrous oxide, and chlorofluorocarbons can have a cumulative effect on climate equal to that of carbon dioxide. A Global Tropospheric Chemistry Program (GTCP) (NRC, 1984) and an International Global Atmospheric Chemistry (IGAC) program (Galbally, 1989) have been proposed to study the sources, transport, reactions, and removal of trace species in the global atmosphere.

The GTCP will measure and model concentrations and distributions of gases and aerosols in the lower atmosphere, chemical reactions among atmospheric constituents, sources and sinks of important trace gases and aerosols, and exchange of gases and aerosols between the troposphere and the biosphere, the earth's surface, including the ocean, and the stratosphere. Activities include field, laboratory, and modeling studies designed to provide a better understanding of the chemical reactions in the lower atmosphere (troposphere) and to develop new instruments for measuring trace atmospheric constituents.

Much of the GTCP will take place in the context of the IGAC program.

IGAC is an initiative of the Commission of Atmospheric Chemistry and Global Pollution of the International Association of Meteorology and Atmospheric Physics of ICSU. It should be noted that neither of these programs is aimed at the carbon dioxide issues. The designation of IGAC as an IGBP core project program follows a decision reached jointly by the Commission and the Special Committee for the IGBP to expand the original scope of IGAC to include a strong biological component dealing with sources and sinks of biogenic gases.

The exchange of trace species between the ocean and the atmosphere is an important boundary process for the ocean. Thus the mechanisms for interaction between these atmospheric chemistry programs and JGOFS and related programs in the ocean must be enhanced. A close working relationship is now developing between JGOFS and IGAC, and plans are being made for a formal relationship between these two programs, which would be particularly appropriate since they are both part of the IGBP.

Oceanic Ecosystem Response to Climatic Change

Global oceanic fluxes of carbon and other materials are mediated through the complex interactions of the growth, reproduction, and mortality dynamics of oceanic communities and their constituent populations. This happens through the consumption of primary production, transfer up the food chain to harvestable resources, recycling of primary nutrients to enable continued primary production, processing of residual material, and its sinking and sediment burial. Biological responses to changes in climate and the implications for carbon flux are important topics that are receiving increased attention today. The initial planning stages of a new program, the Global Ocean Ecosystem Dynamics (GLOBEC) program, are now under way. GLOBEC is aimed at understanding how a changing global environment will alter the stability and productivity of marine ecosystems. Note that the variability in primary production (a vital concern to the JGOFS program) cannot be understood without taking into account the organisms that supply a significant component of nitrogen to the phytoplankton and at the same time control, at certain places and times, phytoplankton abundance, through grazing.

Equally important is the role of the secondary producers of the lower levels of the food chain in the transfer of energy and carbon up the food chain. This transfer has important implications for the animal populations, e.g., fisheries, of vital concern to humans. The need for understanding the physical and chemical oceanography of the upper ocean, where most of the life occurs, is a fundamental aspect of understanding the biological systems.

Most marine species, including zooplankton, bottom-living animals, and fish, base their strategy for long-term survival on the production of hundreds, thousands, and even millions of offspring by every female adult. The implications of this almost universal strategy are obvious. First, survival of the individual in the ocean is already very tenuous, and high rates of mortality are virtually guaranteed. Second, changes of seemingly insignificant percentages in survival (e.g., between 0.01 and 0.001 percent) produce enormous differences in adult numbers and biomass, considering that, for example, fish can grow through 6 orders of magnitude of biomass increase. The central question therefore at the heart of the GLOBEC program is "what is the potential for global-scale climatic change to disturb the already extreme variability inherent in natural ecosystems beyond the point of recovery?" For instance, what will be the fate of coral reefs, estuaries, or major fish stocks?

The GLOBEC planning currently is examining the best ways to address (1) the development of new ecological theory applicable to oceanic ecosystem dynamics, (2) new modeling approaches that can lead to prediction of ecosystem changes, and (3) a new generation of in situ technology to measure populations as they fluctuate in real time in response to the rapidly changing physical environment (NRC, 1987). As these new ideas and techniques are developed, they will be incorporated into an ongoing program and will provide insights into biogeochemical processes.

Physical Processes

The fundamental physical setting of the ocean and the interaction of the ocean with the atmosphere are essential aspects of the chemical and biological interactions. As a consequence, it is essential that such programs as the Tropical Ocean-Global Atmosphere (TOGA) program, WOCE, and the Global Energy and Water Cycle Experiment (GEWEX) be carried out by the World Climate Research Program (WCRP) to provide the necessary description and understanding of the physical processes in the ocean.

Tropical Ocean-Atmosphere Interactions

Considerable progress has been made in the past 2 or 3 years in actually predicting change in the study of El Niño, the periodic anomalous warming (and cooling) that occurs in the tropical Pacific Ocean accompanied by global atmospheric changes. The economic impacts of associated excess rainfall, flooding, and droughts have been estimated in the billions of dollars. To study this phenomenon, the WCRP developed the TOGA program as its first major project in 1985. TOGA has established a quasi-operational monitoring network of drifting and moored buoys, sea level gauges, and upper-layer and meteorological measurements from volunteer observing ships in the tropical Pacific, Atlantic, and Indian oceans. By focusing on the tropical system, we are beginning to learn how the climate system works (NRC, 1990).

Studies are now being carried out to determine the frequency and magnitude of El Niño events in the geologic past, using the sediment record. The information from the sediments also gives a record of biological activity. This statistical data could be useful in understanding the long-period fluctuations and interactions of biology and physics in the El Niño.

Global Ocean Circulation

On longer time scales, from interannual to decadal, we need to understand such questions as how much heat is transported by the ocean and how does the ocean take up and redistribute carbon dioxide and other trace gases important in the radiative balance. To address these issues, the second major ocean project of the WCRP is the World Ocean Circulation Experiment (WOCE). Its primary goal is to develop models useful for predicting climatic change and to collect the data necessary to test them. Specific parts of the program will include efforts to determine and to understand on a global basis the large-scale fluxes of heat and fresh water, their divergences over 5 years, and their annual and interannual variability. WOCE will also try to identify those oceanographic parameters, indices, and fields that are essential for continuing measurements in a climate observing system on decadal time scales and to develop cost-effective techniques suitable for deployment in an ongoing climate observing system (U.S. WOCE Office, 1989; WOCE, 1988).

WOCE has as central observational elements a global observing network of precision satellite measurements of the surface winds and currents, direct current measurements, and precise measurement of temperature, salinity, and chemistry. WOCE will begin its field phase in 1990. The JGOFS program will use WOCE logistics to carry out a global survey of carbon dioxide in the sea. In early 1991 the in situ programs will be supported by the launch of the ESA's ocean satellite ERS-1. ERS-1 will provide global wind measurements by scatterometer and surface topography measurements by altimeter for ocean circulation. In 1992 the joint U.S.-French precision altimeter mission TOPEX/POSEIDON will begin to provide accurate measurements of surface topography. In 1995 the Japanese Advanced Earth Observing Satellite will provide a flight for the NASA scatterometer (NSCAT), and, in the late 1990s, altimetry and scatterometry will be provided by EOS on the polar platforms.

A scatterometer measures the strength and direction of the surface wind on the ocean and thus is of interest for WOCE studies of air-sea fluxes of momentum. The direct effect of the wind is to produce turbulent mixing in the ocean, a physical process that directly affects biological processes. Upwelling in the ocean is related directly to the wind at the surface. Thus flight of a scatterometer will greatly help in the description and interpretation of biological processes. This point underscores the need for better understanding of the upper ocean.

With the various global data sets, modelers are expected to be able for the first time to realistically portray the oceanic circulation and its interaction with the atmosphere on a global scale. This information can be used to help to develop models that include biological and chemical processes.

WOCE will be augmented by the NOAA Atlantic Climate Change program, which will study air-sea interaction in the North Atlantic Ocean. The Atlantic Climate Change program will provide valuable information on the environmental context for biological studies (NOAA, 1990).

For both TOGA and WOCE the primary question from the biogeochemical point of view is whether these programs will provide the physical understanding of the ocean that is needed to meet the objectives of JGOFS and related biogeochemical programs. Close interaction between the planning for these programs and the needs of the biogeochemical studies needs to be maintained. For example, it is clear that better understanding of the upper ocean is required, but it is not clear whether TOGA or WOCE will collect data on the upper ocean that is sufficient to meet the needs of the biogeochemical programs. It will be the responsibility of the scientists involved in programs like JGOFS and GLOBEC to identify what additional physical studies need to be done.

Precipitation over the Oceans

It has long been recognized that the difference between precipitation and evaporation—the flux of fresh water—is one of the factors that influence oceanic circulation and the chemistry and biology of the ocean. Although evaporation can be estimated with some difficulty from sea surface temperature and surface wind (surface humidity is also required, but difficult to measure at sea), precipitation cannot, except at islands and from ships at sea.

The Global Precipitation Climatology Project (GPCP) provides precipitation data from operational satellites. Sponsored by the WCRP, the program incorporates conventional rain gauge measurements for continental areas and satellite images for estimating water content and precipitation. GPCP began operations in 1987 and will provide global precipitation fields for the period from 1986 to 1995 (WOCE, 1988).

The new satellite microwave techniques, which work in the frequency ranges that are sensitive to the presence of liquid water, are already providing measurements of rainfall over both oceans and land. However, previous microwave measurements have all been made from sun-synchronous polar orbit. Because diurnal rainfall variations are known to be large, such data may not yield representative daily rainfall averages. The first scheduled application of the techniques on a global scale from a special tropical, nonsun-synchronous orbit over the oceans will be the Tropical Rainfall Measurement Mission (TRMM), a joint U.S.-Japanese mission planned for the mid-1990s. These measurements will be of importance to understanding biological processes. In the late 1990s, rainfall measurements will be carried out from the EOS polar platforms (NASA, 1986).

Polar Processes

The importance of high-latitude studies needs to be emphasized. The Arctic System Science (ARCSS) program is aimed at understanding the physical, chemical, and biological interactions that link the arctic environment to global climate. If successful, ARCSS will provide improved information and predictive modeling capabilities of physical and biological conditions and changes in the planet's environmentally sensitive polar regions. Using data from ice and sediment cores, ARCSS should help to expand the understanding of arctic paleoenvironments. JGOFS is beginning to include high-latitude process studies in their planning (SCOR, 1990).

Southern polar regions are also important. In its 1989 report *The Role of Antarctica in Global Change: Scientific Priorities for the IGBP*, the Scientific Committee for Antarctic Research (SCAR) noted that the Southern Ocean covers only 10 percent of the world ocean but plays a major role in the global carbon flux. It has a significant influence on the interannual variation of the world ocean's capacity to take up atmospheric carbon dioxide. Much of the world's deep water is formed in the sea ice zone of the Southern Ocean, and high wind stress, local microbial productivity, and sea ice cover all vary to produce a range of potential incorporation of gases in surface waters. The flux of biogenic materials in the Antarctic is linked to the formation of, sinking, and northward movement of cold water as well as to the transport pattern of water masses. Changes in the fluxes are well documented in the upper layers of ocean sediments. JGOFS is now addressing issues of the biogeochemical fluxes in the Southern Ocean, as part of integrated and essential parts of the IGBP in the Antarctic.

One particular emphasis for an antarctic IGBP investigation, noted by the SCAR group, is to study the effects of ocean changes on the dimethylsulfide-emitting phytoplankton, as noted earlier. Abundant unicellular algae in the Southern Ocean are emitters of dimethylsulfide, which, in the atmosphere, may change cloudiness patterns and either enhance or diminish the greenhouse effect.

Another subject of concern is the effect of ozone depletion on antarctic organisms. The subsequent increased levels of ultraviolet radiation could affect the phytoplankton that constitute the base of the food web in aquatic ecosystems. Studies have shown that increased levels of ultraviolet exposure result in reduced primary production and an altered community structure.

By weakening the base of the food web and altering trophodynamic relationships, ultraviolet-induced changes could affect the entire Southern Ocean ecosystem.

In terms of measurements in the polar regions, satellite-borne instruments are invaluable. Currently, measurements are taken from operational satellites such as the Defense Meteorological Satellite Program (DMSP), which makes microwave measurements of snow and ice. The European ERS-1 and the Japanese ERS-1 satellites will provide polar snow and ice measurements until EOS is in place in the late 1990s.

STATUS OF MODELING AND MONITORING EFFORTS

The Need for Modeling

As indicated above, there are several large research programs either in progress or planned for the early 1990s to study the role of the ocean in climatic change. Moreover, the technology necessary for improving the speed of computers to handle global ocean prediction models is developing rapidly. It appears that the next generation of supercomputers, relying on high-speed parallel processors and other new developments, will provide the necessary number-crunching needed to incorporate the oceans in long-term studies of climate in a physically realistic way.

Two kinds of models are required: (1) those that simulate the existing knowledge (diagnostic) and (2) those that try to develop a better understanding of the world (predictive). Both kinds of modeling are carried out in these programs. A prerequisite for an efficient monitoring scheme that covers the broad spectrum of physical and biogeochemical processes is to have models of the way these processes work. Models of the physical processes and of primary production processes are being developed through TOGA, WOCE, and JGOFS. Development of models of secondary production is a major raison d'etre for GLOBEC.

The Need for Monitoring

A major piece of a global effort to understand the linked physical and biogeochemical systems is still missing: it is a routine, global, operational ocean-observing system that monitors physical, chemical, and biological parameters. Such a system must be put into place if we are to describe, understand, and ultimately predict global change. For understanding biogeochemical fluxes, it is especially important to monitor the dissolved gases such as carbon dioxide.

For the atmosphere, we have the World Weather Watch (WWW), which consists of a combination of satellite and in situ measurements in the atmosphere. Each participating nation has a national weather service that provides local data for transmission on the Global Telecommunications System (GTS). The worldwide satellite network, consisting of five geostationary satellites operated by the United States, ESA, Japan, and India, and polar-orbiting satellites operated by the United States and the Soviet Union, also provides its data through the GTS. This operational system is the basis for the World Weather Watch. It is driven by the customer needs of weather forecasting and civil aviation.

But there is no equivalent system for long-term systematic oceanic observations, primarily because the same level of customer interest has not existed. Most countries do not have the ocean equivalent of a weather bureau, and those that do, like the United States, do not provide the necessary funding to make it viable. In the United States the National Ocean Service of NOAA has the charge for long-term observations, but there has never been a sufficient federal funding commitment to make it work. We do have a pilot monitoring scheme for the tropical Pacific Ocean as part of the TOGA program. The data from this system have been valuable in helping TOGA scientists develop an operational ocean model for El Niño predictions.

The international framework is in place, through the International Global Ocean Station System (IGOSS), jointly sponsored by the Intergovernmental Oceanographic Commission and the World Meteorological Organization. IGOSS supports a global system of expendable bathythermograph (XBT) and related measurements in the upper ocean from volunteer observing ships; the data are transmitted to data centers by the GTS. But in the main, the funds for these XBTs come from research programs like TOGA and WOCE. If we are to see a long-term operational system, then we must find a way to provide such instruments on a regular basis outside research funding. And we must extend the measurements to include new techniques such as acoustics and a wider set of physical, chemical, and biological parameters.

A beginning has been made in the monitoring of dissolved carbon dioxide on a global scale by JGOFS, and there has been monitoring of biological processes in the sea. For example, the monitoring of fisheries stocks sponsored by the International Council for the Exploration of the Sea (ICES) in the North Atlantic since the turn of the century still continues. Regular monitoring of biological parameters has been carried out by the CALCOFI program off the coast of California. Stations like Station Papa in the northwest Pacific Ocean have carried out biological monitoring. The U.K. continuous plankton recorder has been operated in the waters around the U.K. and all across the Atlantic Ocean. Ocean color was monitored for several years by the now-inoperable Coastal Zone Color Scanner on the Nimbus-7 satellite, but as noted above in the section on JGOFS, we are facing a long delay in ocean color monitoring now.

In the Antarctic, the SCAR group gave emphasis to monitoring changes in organisms and biological processes that are directly linked to environmental changes. Phytoplankton, at the base of the marine food web, may be more useful to monitor than the top marine predators, which can be affected by a greater complexity of events. However, the uppermost predators, such as seals, integrate changes over several seasons and may be better indicators of long-term trends. Historical and cohort strength fluctuations in several species of antarctic seals may be related to fluctuations in environmental parameters such as pack-ice extent, krill, and even El Niño events. Corals and molluscs show annual growth increments, and demographic analysis of populations may show the integrated effects of past changes.

But this is only a start compared to the network of measurements that is required. Understanding global change requires global measurements in both the atmosphere and the ocean; for the long term, we will require operational measurements. This transition from research to operations must be a focus for the 1990s and into the twenty-first century.

Satellite measurements are essential to monitoring: for the first part of the 1990s the various research missions that have been described above will provide the necessary information. But for the late 1990s and beyond, for true monitoring of global change, instruments of EOS need to be in place to monitor ocean color, wind stress, and ocean currents, as well as ocean temperature and rainfall. EOS is now scheduled to be launched in 1998. It is important that the various research missions and systems proposed be able to provide measurements until the EOS is in place, so that continuity of data is provided. If there are delays in EOS, then ways to extend the research missions should be considered.

RECOMMENDATIONS FOR ENHANCED SUPPORT, NEW INITIATIVES, AND RESEARCH PROGRAMS

Based on the discussion above, the committee recommends the following:

- Support for JGOFS as a core program of the IGBP. The JGOFS carbon dioxide survey, modeling, and process study components should be supported at the required levels. Ocean color by satellite is of particular importance.
- Biogeochemical studies in the coastal regions. JGOFS, primarily aimed at the open ocean, will not address all of the issues related to biogeochemical dynamics in the ocean. Of special interest are the coastal regions. It is in these regions that much of the biological productivity takes place, and yet it has been difficult to define a program because of the complexity of processes there. The committee believes that this should be a focus for the next phase of planning. Nationally, such planning should be part of new coastal oceanography programs. Internationally, such planning for studies

of interdisciplinary coastal processes is now a part of IOC and IGBP planning.

• Chemical fluxes across the sea surface. It is important that the JGOFS program continue to work closely with the relevant atmospheric chemistry programs, in particular IGAC, so that fluxes between the atmosphere and the ocean are understood and the boundary conditions are established. Increased cooperation and coordination between JGOFS and IGAC are encouraged and should be strengthened.

• Ecosystems dynamics. It is important to recognize explicitly that the role of the biota in global change in the ocean is not limited to its mediation of biogeochemical cycles. Research must be conducted on the role of global climatic change on the production, ecosystem structure, and fate of populations vital to the health and continued existence of humankind.

• Physical studies of the ocean. The fundamental physical setting of the ocean and its interaction with the atmosphere are essential aspects of the chemical and biological interactions. As a consequence, it is essential that programs such as TOGA and WOCE be fully funded so they can be carried out successfully. Enhanced interactions and project planning between biogeochemical and physical programs are needed to ensure that the physical understanding of the ocean that is needed to meet the objectives of JGOFS and related biogeochemical programs is provided.

• Upper ocean physics and chemistry. Neither TOGA nor WOCE is designed to provide the detailed physical knowledge of the upper ocean needed for full understanding of biological processes and their changes. The JGOFS and GLOBEC programs will need to identify clearly what is required to address these issues.

• New measurement techniques. Ocean color measurements in the near and far term are essential. The SEAWIFS program needs to be supported, as does the TOPEX/POSEIDON altimeter, the NASA scatterometer, and the TRMM mission. Data from the European ERS-1 satellite should be fully exploited. Finally, full support is needed for the flight of the ocean-related instruments on EOS, now scheduled for 1998 and beyond. If there are delays in the implementation of EOS, then ways to extend the proposed research missions must be found in order to provide continuity of data. All of these instruments are of crucial importance in describing and understanding biological processes in the ocean.

• Modeling. With the various global data sets from the scientific programs and from operational monitoring, modelers are expected to be able for the first time to realistically portray the ocean circulation and its interaction with the atmosphere on a global scale. This information can be used to help to develop models that include biological and chemical processes.

• Monitoring. A major piece of a global effort to understand the linked physical and biogeochemical systems is still missing: it is a routine, global, operational ocean-observing system that monitors physical, chemical, and biological parameters. If a long-term operational system is to be in place, ways must be found to provide such instruments on a regular basis outside research funding. The international coordination mechanisms, such as the IOC, could play a strong role. This transition from research to operations must be a focus for the 1990s and into the twenty-first century.

REFERENCES

Charlson, R.J., J.E. Lovelock, M.O. Andreae, and S.E. Warren. 1987. Oceanic phytoplankton, atmospheric sulphur, cloud albedo, and climate. Nature 326:6555-6661.

Galbally, I. (ed.). 1989. International Global Atmospheric Chemistry (IGAC) Program: A Core Project of the International Geosphere-Biosphere Program. IAMAP Commission on Atmospheric Chemistry and Global Pollution. Renwick Pride Pty Ltd., Albury, Australia. 55 pp.

Martin, J.H., R.M. Gordon, and S.E. Fitzwater. 1990. Iron in Antarctic waters. Nature 345:156-158.

National Aeronautics and Space Administration (NASA). 1986. From Pattern to Process: The Strategy of the Earth Observing System. EOS Science Steering Committee Report. Vol. II. NASA, Washington, D.C.

National Oceanic and Atmospheric Administration (NOAA). 1990. Atlantic Climate Change Program: Science Plan. Draft report. NOAA, Rockville, Md. 29 pp.

National Research Council (NRC). 1984. Global Tropospheric Chemistry: A Plan for Action. National Academy Press, Washington, D.C.

National Research Council (NRC). 1987. Recruitment Processes and Ecosystem Structure of the Sea: A Report of a Workshop. National Academy Press, Washington, D.C. 44 pp.

National Research Council (NRC). 1988. Toward an Understanding of Global Change: Initial Priorities for U.S. Contributions to the International Geosphere-Biosphere Program. National Academy Press, Washington, D.C.

National Research Council (NRC). 1990. TOGA: A Review of Progress and Future Opportunities. National Academy Press, Washington, D.C.

Scientific Committee on Antarctic Research (SCAR). 1989. The Role of Antarctica in Global Change: Scientific Priorities for the IGBP. Prepared by the SCAR Steering Committee for the IGBP. ICSU Press, Cambridge, U.K. 28 pp.

Scientific Committee on Oceanic Research (SCOR). 1987. The Joint Global Ocean Flux Study: Background, Goals, Organization, and Next Steps. International Council of Scientific Unions, Paris. 42 pp.

Scientific Committee on Oceanic Research (SCOR). 1990. The JGOFS Science Plan. JGOFS/SCOR, Halifax, Canada, in press.

U.S. World Ocean Circulation Experiment (WOCE) Office. 1989. U.S. WOCE Implementation Plan. U.S. WOCE Implementation Report No. 1. U.S. WOCE Office, College Station, Tex. 176 pp.

World Ocean Circulation Experiment (WOCE). 1988. WOCE Implementation Plan. WCRP Series Vol. I and II, WCRP-11 and WCRP-12. WMO/TD No. 242 and 243. Geneva.

8

Documenting Global Change

OVERVIEW

Two aspects of the U.S. Global Change Research Program will play a particularly crucial role in the success of all aspects of the program: (1) monitoring of the earth system over years to decades in order to document the global changes and (2) information management to make such documentation feasible and to provide information for the various process studies and modeling efforts described in chapters 2 through 7 of this report.

An integral part of developing a monitoring and information management strategy is the identification of how the needed tasks will be accomplished, particularly in the case of satellite observations, where the expense and lead time require extraordinary care in scientific justification and realistic planning. Key missions in this regard include the ongoing NOAA polar orbiting and geostationary satellites, the Department of Defense GEOSAT and ongoing Defense Meteorological Satellite Program (DMSP) series, the Earth Observation Satellite (EOSAT) Landsat series, approved NASA missions, such as the Upper Atmosphere Research Satellite (UARS) and TOPEX/POSEIDON (the U.S.-French ocean topography experiment), and two planned NASA series that are part of Mission to Planet Earth, the Earth Probes and the Earth Observing System (EOS). In addition, a number of other missions are operated by other nations, including the Soviet Union, Japan, France, and

---

This chapter was prepared for the Committee on Global Change by S. Ichtiaque Rasool, Jet Propulsion Laboratory; D. James Baker, Jr., Joint Oceanographic Institutions; and Ferris Webster, University of Delaware; with input from Francis P. Bretherton, University of Wisconsin.

the European Space Agency (ESA). Japan and ESA are partners with NASA in EOS, which will begin to operate in the late 1990s. EOS is an integrated observation and information system with the potential to provide a new generation of capability for understanding and monitoring the earth system.

For the purposes of this chapter, monitoring is defined as the minimal subset of currently achievable sustained global measurements, including their processing to deliverable products, that will document the baseline state of the planet earth and global changes. Information management, which is an extension of data management, is defined as including a comprehensive process of compilation, distribution, and preservation of basic data, derived products, and information about them and is approached in terms of the inputs to and outputs from a variety of scientific activities within global change research.

MEASUREMENT STRATEGY

Monitoring Requirements

Public and scientific concern with global change centers on, but is not confined to, significant changes in the earth's climate in the decades to come, due to increases in atmospheric carbon dioxide concentrations primarily from the increasing worldwide consumption of fossil fuel and to the emission into the atmosphere of other greenhouse gases such as methane, nitrous oxide, and chlorofluorocarbons. The stratospheric ozone layer is also a focus of attention. Other impacts of human activities, such as acid rain, deforestation, and soil degradation, are affecting the global environment in ways that we are only beginning to comprehend and yet surely include interaction with climate and the stratosphere. A significant element in these concerns is the realization that by the time the serious threats to humanity posed by these changes become obvious the changes may be irreversible, at least for several centuries. Furthermore, the driving forces are so deep-seated in our industrialized society and growing world population that mechanisms to control them will be difficult to put in place. Finally, our understanding of the earth system processes is quite inadequate for effective management on the scale that will be required.

The focus here is on the initial scientific design of a monitoring system with an emphasis on those aspects that are already under way or could be implemented in the near future, at least in prototype form. Because of the very nature of the long-term commitment required for monitoring, special institutional and funding arrangements are necessary, major resources are involved, and great selectivity is required. Therefore, for each variable, consideration is given to the following questions:

• Which measurements are both critical to the integrity of the program and feasible for immediate implementation on a long-term basis within the context of existing capabilities and the research programs and process studies currently being planned?

• What is the current status of the monitoring system and what needs to be done to ensure consistency and continuity?

Two kinds of global-scale long-term monitoring are needed: (1) monitoring of the magnitude of the driving forces that may bring long-term changes in the equilibrium state of the earth system and (2) monitoring of the state variables or the "vital signs" of the earth where such changes are liable to manifest.

A comprehensive approach leads to a long list of global measurements that need to be made on an ongoing basis (e.g., Earth System Sciences Committee, 1988, Table 9.1A). In the context of the overall program, relative priorities for individual variables must be judged not only in terms of their contribution to the monitoring requirements but also in view of their importance for validating models and advancing our understanding of specific processes, the magnitude of the effort required, and the state of readiness of the observation and analysis techniques involved.

Table 8.1 is an attempt to identify priority requirements and comment on the current status of each. In many cases, new approaches are being developed. It is clear that the measurement system will evolve with time.

Global Synthesis

In chapter 2, a new framework for earth system modeling is proposed. In order to eventually realize fully coupled, dynamical models of the earth system, a step-by-step approach is required to develop several partial models representing the interface between the terrestrial biosphere and the atmosphere, the coupling of physics and chemistry within the atmosphere, and the interface between the oceans and the atmosphere, including the chemical exchange and the biological dynamics within the upper layers of the oceans.

In order to build these models, we will need time-dependent data sets, often global in extent, which will be used as input to these models and also to test the predictions of these models. These data will be derived by indirect or surrogate global-scale measurements, together with regional- and local-scale process studies, to infer global-scale values of the desired variable.

The need for such data sets, specific for global change studies and the development of realistic earth system models, puts new and stringent requirements on the global observing system. These requirements are simultaneous observation from satellites of several parameters covering large areas over long time periods, subsatellite area coverage and field measurements, global surface observation networks on the land and in the oceans, and subsurface measurements.

TABLE 8.1 Earth System Monitoring

Parameter	Status and Comments[a]
Driving Forces for Global Change	
Solar irradiance	Ongoing; need intersatellite calibration
Solar ultraviolet spectrum	Ongoing from SBUV/Nimbus, UARS (1992 and beyond); need check on long-term consistency
Volcanic aerosols	Ongoing (SAGE) for polar stratosphere, ad hoc measurements from surface observatories; need global monitoring program
Trace gases	
Carbon dioxide	Good coverage in time and space from surface network, NOAA/GMCC
Methane, nitrous oxide, carbon monoxide	Ongoing from BAPMON and ad hoc coverage, field experiments (e.g., ABLE); need reliability, IGAC program should resolve deficiencies
Chlorofluorocarbons	Ongoing from industry statistics, polar ozone expedition
Biomass emission rates of trace gases	Very poor, spotty coverage; ad hoc measurements, NOAA/AVHRR and field experiments may provide global estimates
Land use change	Poor; IGBP initiative (See Report No. 8), NOAA/AVHRR with Landsat and SPOT, EOS/MODIS, HIRIS provide potential measurements
Global Change Symptoms (Vital Signs)	
Global tropospheric temperature	Ongoing from radiosondes, NOAA/TOVS; long-term consistency in coverage and sensor stability and intersatellite consistency are the major issues
Surface temperature	Ongoing from surface meteorological network, satellites for sea surface temperature; issues related to spotty coverage in the southern hemisphere, land and ocean integrated data analysis need to be resolved
Total ozone	Ongoing from Dobson and satellites; need to pay attention to potential gaps in coverage by TOMS
Stratospheric temperature	Ongoing from radiosondes, Nimbus, NOAA, UARS, EOS

TABLE 8.1 continues

TABLE 8.1 (continued)

Parameter	Status and Comments[a]
Upper troposphere water vapor	Ongoing from WWW, GOES, Meteosat and planned for HIRIS/EOS; long-term trends not yet discernible
Clouds	Ongoing from ISCCP data, earth radiation budget, field experiments (FIRE)
Interannual air-sea interaction fluctuation (El Niño)	Ongoing from TOGA data sets; intensive research activity ongoing
Oceanic and atmospheric heat transport and storage	Ongoing and planned from WOCE, JGOFS, NOAA/AVHRR, TOPEX/POSEIDON, ERS 1 NSCATT, EOS; research ongoing
Oceanic carbon dioxide uptake	Ongoing and planned from JGOFS (1989-1996), Nimbus, SEAWIFS, ADEOS, EOS/MODIS
Sea ice extent	Ongoing and planned from Nimbus, DMSP, SSM/I, ERS 1, EOS, JERS 1, Radarsat
Rainfall	Poor; WCRP Global Precipitation Climatology Project, GEWEX, TRMM, EOS, BEST provide potential measurements
Sea level	Ongoing from global network of in situ gauges, satellite altimetry, VLBI
Biospheric parameters/land and oceans	Spotty, poor; measurements from JGOFS, Nimbus 7, NOAA/AVHRR, SPOT, Landsat, Radarsat, ADEOS, EOS, JERS; surface measurements from UN/MAB, NSF/LTER, USFS/CFI
Soils	Very poor; need coordinated surface observation networks, potential expansion of SOTER data base, EOS/SAR

[a] Acronyms and abbreviations used in this table are as follows:

ABLE	Atmosphere Boundary Layer Experiment
ADEOS	Advanced Earth Observing Satellite
AVHRR	Advanced Very High Resolution Radiometer
BAPMON	Background Air Pollution Monitoring Network
BEST	Bilan Energetique de la Système Tropical
CFI	Continuous Forest Inventory

TABLE 8.1 continues

TABLE 8.1 (continued)

DMSP	Defense Meteorological Satellite Program
EOS	Earth Observing System
ERS 1	European Space Agency's remote sensing satellite
FIRE	First ISCCP Regional Experiment
GEWEX	Global Energy and Water Cycle Experiment
GMCC	Global Monitoring for Climate Change
GOES	Geostationary Operational Environmental Satellite
HIRIS	High-Resolution Imaging Spectrometer
IGAC	International Global Atmospheric Chemistry Program
IGBP	International Geosphere-Biosphere Program
ISCCP	International Satellite Cloud Climatology Project
JERS 1	Japanese Earth Resources Satellite
JGOFS	Joint Global Ocean Flux Study
LTER	Long Term Ecological Research
MAB	Man and the Biosphere
MODIS	Moderate Resolution Imaging Spectrometer
NOAA	National Oceanic and Atmospheric Administration
NSCATT	NASA Scatterometer on ASEOS
NSF	National Science Foundation
SAGE	Stratospheric Aerosol and Gas Experiment
SAR	Synthetic aperture radar
SBUV	Solar Backscatter Ultraviolet
SEAWIFS	Sea-viewing, Wide Field-of-View Sensor
SOTER	Soil Terrain Digital Data Base at Scale 1:1m
SPOT	Système Probatoire de l'Observation de la Terre
SSM/I	Special Sensor Microwave/Imager
TOGA	Tropical Oceans and Global Atmosphere Program
TOMS	Total Ozone Mapping Spectrometer
TOPEX/POSEIDON	U.S.-French Ocean Topography Experiment
TOVS	TIROS Operational Vertical Sounder
TRMM	Tropical Rainfall Measuring Mission
UARS	Upper Atmosphere Research Satellite
UN	United Nations
USFS	U.S. Forest Service
VLBI	Very Long Baseline Interferometry
WCRP	World Climate Research Program
WOCE	World Ocean Circulation Experiment
WWW	World Weather Watch

TABLE 8.2 Examples of Globally Synthesized Products Derived from Earth System Measurements

Product	Derived from
Latent heat flux	Radiation balance, temperature, moisture, field studies
Global surface air temperature	Surface radiating temperature
Vegetation type, plant stress	Land cover change, greenness index
Ocean chlorophyll	Ocean color
Oceanic uptake of carbon dioxide	Ocean temperature, surface wind, ocean color, atmospheric carbon dioxide
Oceanic and atmospheric transport	Wind, temperature (ocean and heat atmosphere), ocean currents
Trace gas emission from biomass burning	Fire frequency and intensity, regional trace gas concentrations

From an observing system consisting of the elements above, we can begin to derive globally synthesized products such as those in Table 8.2.

In addition to new data that will need to be collected and synthesized on a global scale, existing data could provide useful information if made accessible to the research community. Data sets classified for military intelligence purposes, such as the global data set on digital terrain information important for many aspects of global change research ranging from surface energy interactions to surface roughness fields for circulation models, could provide valuable information if released.

Process Studies

Chapters 3 through 7 of this report identify the observation needs that must be given priority to make progress in each field of research. It is clear that the pace of activities in each of these areas is largely limited by the available data. This section summarizes data needs by providing examples for each chapter to provide an understanding of the scope of the observation and monitoring program required to implement a research program.

Earth System History and Modeling

The geologic record is the only source of information on how the climate system has evolved through time, and in chapter 3 the specific important geoscience contributions to global change research are enumerated, along with the observational needs for sustaining research in this area. A global data base of paleoclimate observations is needed and will draw on a great diversity of paleoenvironmental sensors, including direct observations, historical documents, anthropological records, tree rings, ice cores, lake and ocean sediments, and corals.

Measurements are also needed to quantify observed environmental changes in terms of temperature and precipitation. For example, because they have great significance for human activities such as food production, additional high-resolution marine records are critically needed to support regional process studies. Multiple independent monitors of changes in temperature, precipitation, biota on land, dust and sulfate aerosols in the atmosphere, and atmospheric concentrations of carbon dioxide and methane are also needed. More measurements of sea surface temperature, deep and intermediate waters, aeolian fluxes, and components of the carbon cycle are also needed.

Fractionation of isotopes in precipitation, plankton, and tree rings; entrapment of gases within ice; and incorporation of trace metals into corals are the types of process studies of modern environments that would advance our knowledge of global change.

Human Sources of Global Change

Chapter 4 formulates a research plan for achieving a better understanding of the human sources of global change. The process studies necessary to accomplish this goal will need measurements of the amounts of energy and materials being used per unit value of production, population density, economic activity, and land tenure pattern.

Data collection in this area has both a historical and a current component. It includes data on human activities that lead to changes in the chemical flow, physical properties, and surface covers of interest as well as data on demographic, technical, and socioeconomic variables.

Collection of these types of data is complicated by the fact that many of the coefficients for industrial processes, such as carbon dioxide emission coefficients for various energy technologies, are well documented, whereas coefficients for land use processes, such as methane from various rice cultivation techniques, are not. Better data collection strategies need to be developed, especially in the area of global land cover change and its impact on global climate.

Water-Energy-Vegetation Interactions

There have been few successful efforts, either in modeling or in data acquisition, to link the activities addressing the physical climate with those addressing the terrestrial biosphere so as to further understand and improve our capability to predict global change.

Chapter 5 focuses on the interactions between the vegetated land surface and the atmosphere, particularly on the exchanges of energy, heat, and carbon dioxide between the two. This goal requires development of comprehensive biophysically based models of the atmosphere and land biosphere that will utilize measurable parameters.

Table 5.3 identifies the ongoing field campaigns currently providing the regional data measurements and lists the campaigns being planned to extend the data collection process to meet the research requirements. The parameters of interest being sought by these campaigns are referenced in the tables.

Terrestrial Trace Gas and Nutrient Fluxes

Although the importance of photosynthesis and respiration in controlling carbon dioxide and oxygen has long been known, the biospheric processes controlling nitrogenous compounds such as nitrous oxide, nitric oxide, and ammonia, sulfur compounds such as hydrogen sulfide, and various hydrocarbons have only recently been appreciated.

Chapter 6 investigates this recent development and places it in the context of our current environment, where, for the first time in the history of the earth, these natural and human-caused atmospheric and biospheric processes may alter the global climate with potential impacts on human welfare.

The data needs of this research are associated with process studies that relate methane production, consumption, and flux to environmental parameters such as burning and livestock farming and to changes in ecosystem structure and function. The dynamics of collecting these data must emphasize the integration of information obtained at different scales from simultaneous chamber, tower, and aircraft flux measurements. Also, better spatial and temporal coverage of atmospheric methane concentrations and isotopic composition (carbon and hydrogen) in source regions must be obtained. Field tests and models also need to be developed relating fluxes of water, sediment, nutrients, and pollutants to interactive fluxes of trace gases between the biosphere and the atmosphere. Table 6.1 summarizes the environmental variables regulating the fluxes of trace gases from terrestrial ecosystems, giving a flavor of the complexity of the observational requirements of this research.

Biogeochemical Dynamics in the Ocean

Chapter 7 identifies the efforts, including those currently under way, required to develop the capability to predict the effect of projected climatic change on the ocean's physical, chemical, and biogeochemical processes, especially as they feed back to climate via the release and absorption of radiatively important gases. The chapter gives the status of ongoing and proposed programs that focus on five areas of investigation and their attendant data measurement needs: (1) biogeochemical flux with emphasis on carbon; (2) ocean-atmosphere interface; (3) oceanic ecosystem response to climatic change; (4) underlying physical processes in the oceans and atmosphere; and (5) processes in the polar regions. The data needs are global and long term and will involve satellites, surface buoys, and subsurface sounding to assess a range of parameters, including sea surface temperature, radiation balance state, winds, chlorophyll, atmospheric carbon dioxide, oceanic carbon dioxide, subsurface dynamics, and ocean topography. Together these data will be used to estimate fluxes of energy and trace gas at the ocean-atmosphere interface.

Existing and Planned Observing Systems

In order to meet the data requirements for documenting global change, for developing and testing models at the interfaces of land, oceans, and atmosphere, and for undertaking continental-scale process studies, a measurement program that has the following elements is needed:

- a satellite system for measuring a number of parameters, often simultaneously, with a time scale ranging from seconds to decades and a space scale ranging from pixels to global;
- large-scale field and process studies involving satellites, aircraft, balloon, and surface observing stations;
- a global observing network on the earth's surface for measuring those variables that cannot be observed from space and for validating and calibrating the remotely sensed measurement; and
- two modeling activities, one to help decide the optimal design of the monitoring system and the second to derive data products from indirect and surrogate measurements.

Space Observing System

The science requirements are developed in the preceding chapters in a context of ever-improving techniques for global observations, many of which are dependent on satellites for the global, synoptic, and long-term view. The current international operational satellite system meets some of the science requirements, but it is clear that it could be upgraded and expanded with existing technology to produce many of the long-term data that will be needed for a program to study global change.

In order to upgrade the system, it will be necessary to use the technology that has been developed on specialized research missions, to carry out data validation experiments and to establish new comprehensive data archives and data dissemination systems. Such an upgrade is indeed feasible if the proper support is provided. Therefore the current satellite system, augmented with technology developed by research missions and supported by validation experiments and a comprehensive data system, could provide the basis for a global change observing system.

In order to actually develop the current system into a system for the study of global change, a certain class of actions needs to be taken immediately. These include validation of current data sets, transfer of demonstrated new technology to operations, identification and filling of gaps in the system, and finally developing a "total" system, such as EOS, that will carry out the space observation program for the next several decades.

To establish the necessary parts of the global change observing system, we need to look first at existing activities. Current NOAA, ESA, Japanese, and Indian operational satellites produce routine data products on a number of parameters important to global change. These include cloud cover, sea surface temperature, atmospheric temperature profiles, vegetation index, and ice cover. The study of global change requires that these measurements be continued and at the same time adequately validated. The World Climate Research Program (WCRP) has started to produce long-term validated data sets for climate purposes, including cloud climatology, sea surface temperature, radiation budget, precipitation, surface winds, and ocean currents. At the same time, the International Satellite Land Surface Climatology Program (ISLSCP) is planning to produce validated data sets on surface albedo, land surface temperature, vegetation cover, and evaporation and transpiration.

It is therefore important that the necessary support be provided to complete the validation experiments of existing programs and to provide for the selection of appropriate algorithms to produce routinely the data sets crucial to studies of global change. A number of research missions flown during the past decade have shown that it is feasible to measure these critical parameters on a global scale. The Nimbus series, Seasat, the Geostationary Operational Environmental Satellite (GOES), and Shuttle-based tests of instruments have provided valuable information on how to measure properties of the earth ranging from the radiation budget to ocean primary productivity. Perhaps the best example of using satellite measurements for constructing large-scale data bases is found in the approval and initial stages of the implementation of ESA's remote sensing satellite, ERS-1. This satellite system is the first of an operational series of satellites aimed at an integrated set of measurements of the ocean, land, and atmosphere. The ERS-1 design includes a full validation program and a data system, making it potentially a good model for larger systems aimed at studying global change.

Other examples include NASA's Upper Atmosphere Research Satellite (UARS), which is designed to study the chemistry, radiation, and dynamics of the stratosphere; the joint U.S.-French TOPEX/POSEIDON precision altimetry mission to observe global ocean currents; and the Japanese ocean and land observing program involving an Advanced Earth Observing Satellite (ADEOS) platform and the Japanese Earth Resources Satellite (JERS). (For additional examples, see Figure 8.1.)

An important concern in the near term is the discontinuity of key measurements such as global stratospheric ozone levels, the earth's radiation budget, and the biological productivity of the oceans, made by satellite missions launched in the 1980s. To bridge these gaps in data sets, special attention has to be paid to ensure (1) regular launches of Total Atmospheric

DOCUMENTING GLOBAL CHANGE   227

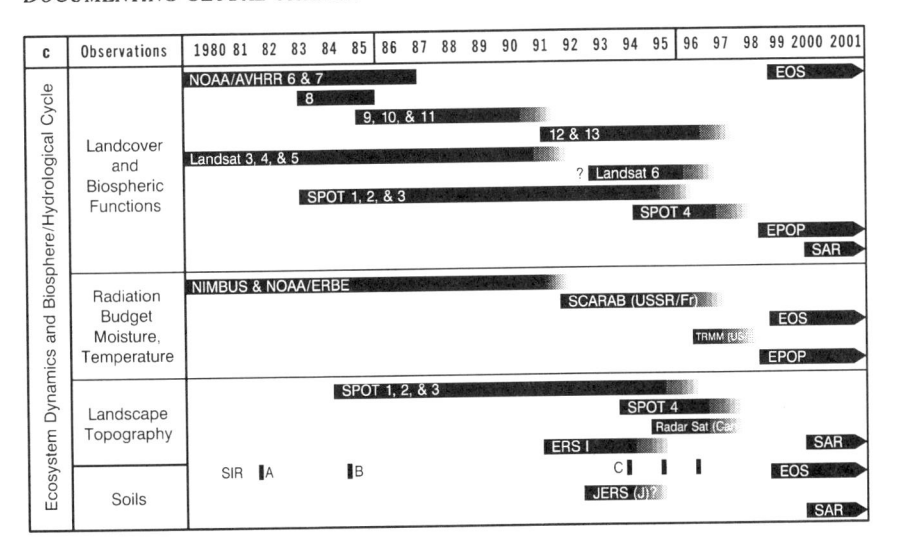

FIGURE 8.1 Space systems for data product continuity for global change studies. (a) Biosphere-atmosphere interactions. (b) Global ocean flux study. (c) Ecosystem dynamics and biosphere-hydrological cycle. (Reprinted from International Geosphere-Biosphere Programs (1990). Copyright © by IGBP.) NOTE: Acronyms and abbreviations used in this figure are as follows: ADEOS, Advanced Earth Observation Satellite; ATSR, Along-Track Scanning Radiometer; AVHRR, Advanced Very High-Resolution Radiometer; CZCS, Coastal Zone Color Scanner; EOS, Earth Observation Satellite System; EPOP, European Polar Orbiting Platform; ERBE, Earth Radiation Budget Experiment; ERS, ESA Remote Sensing Satellite; ESA, European Space Agency; GEOSAT, Geodetic Satellite; JERS, Japanese Earth Resources Satellite; JPOP, Japanese Polar Orbiting Platform; MAPS, Measurement of Air Pollution from Space; NOAA, National Oceanic and Atmospheric Administration; NSCATT, Navy Scatterometer; OCTS, Ocean Color Temperature Scanner; SAR, synthetic aperture radar; SCARAB, Scanner for Radiative Budget; SEAWIFS, Sea-viewing, Wide Field-of-View Sensor; SIR, Shuttle Imaging Radar; SPOT, Système Probatoire de l'Observation de la Terre; TOPEX/Poseidon, U.S.-French Ocean Topography Experiment; TRMM, Tropical Rainfall Measuring Mission.

Ozone Sounds (TOMS), (2) intercalibration of Scanner for Radiative Budget (SCARAB) with those sensors on NOAA's Earth Radiation Budget Experiment (ERBE) and the Cloud and Earth's Radiant Energy System (CERES) on EOS, and (3) earliest possible launch of the Sea-viewing Wide Field-of-View Sensor (SEAWIFS), perhaps as early as 1993, and to make sure that continuity of compatible data is maintained with the launch of Japan's Ocean Color Temperature Scanner (OCTS) on ADEOS in 1995.

Even with the extensive operational system that can be planned with existing technology and with the flight of ERS-1 and other planned missions, there will still be major gaps in our ability to measure several other critical parameters identified in other chapters of this report. Of particular importance are direct measurements of precipitation, soil moisture, certain tropospheric trace gases, and aerosols. Therefore, although the WCRP is collecting precipitation data sets with existing techniques, we need research missions to test technology for direct measurement of precipitation and soil moisture. Measurements from instruments such as NASA's Tropical Rainfall Measuring Mission (TRMM) and the French mission BEST will be important for such tests.

In the longer term, that is, from the mid-1990s to the first decade of the twenty-first century, we look to new space platforms with improved instrumentation and techniques, closely coupled with ground-based networks and an international data management system. Polar platforms and related satellites defined by NASA, NOAA, ESA, and Japan should include a core payload for simultaneous measurement of atmospheric dynamics and composition, land surface properties, and oceanic parameters. Of particular importance to studies of global change will be polar platform measurements of vegetation characteristics, ocean color, stratospheric parameters, ice and snow cover, and components of the earth's radiation budget. A long-term international commitment to the polar platform will be required for long-term continuity of these data sets.

The proposed EOS program is intended as a major advancement in the science and technology of global remote sensing and includes international contributions from the European and Japanese space agencies. The program promises to integrate a number of related—but previously unavailable or disparate—space-based measurements into one continuing system and to greatly enhance research capabilities, while providing a test bed for the development of the next generation of operational, earth-observing instruments and measurement techniques. Consistent with the needs of the USGCRP, EOS is designed to yield a long-term, continuous set of high-priority measurements on a global basis. The EOS plans call for measurements to be combined for the first time with other important data obtained from space-, suborbital-, and surface-based sources into the EOS Data and Information System (EOSDIS). The comprehensive contents of the EOSDIS should enable the scientific community to document, monitor, and model environmental change, to broaden our understanding of the entire earth system, and to improve predictive capabilities. Both the EOS spacecraft and EOSDIS are planned as interdisciplinary, interagency, international endeavors, all of which are essential features of the USGCRP.

Large-Scale Field and Process Studies

As part of an integrated measurement program, it will be necessary to undertake continental-scale field and process studies. These experiments will represent the culmination of the development of methodologies for deriving quantitative information concerning land surface climatological variables from satellite observation of the radiation reflected and emitted by the earth. Such efforts require the cooperation of researchers working in the fields of remote sensing, atmospheric physics, meteorology, and biology and so are interdisciplinary in nature.

To achieve their objectives, active efforts will be made to acquire data over a range of spatial scales. These data will be used to test various methods of integrating our understanding of small-scale processes up to the scale of satellite pixels of various resolutions. The focus of the experiments then is bound directly to the problem of studying processes and states over a range of scales from individual plant leaves up to fluxes over the entire experimental site.

The First ISLSCP Field Experiment (FIFE) is one of several of these studies that are already being undertaken. Figure 8.2 describes the experiment in graphical terms, showing the levels of interaction and cooperation required to successfully achieve its objectives.

The Hydrologic Atmospheric Pilot Experiment (HAPEX) is another project looking at processes on a larger scale and involves the following elements:

• The study of the process of evaporation and the energy and water balance of the region with a view to developing parameterizations of regional-scale fluxes. It will be necessary to determine how radiation energy is intercepted by vegetated surfaces and also to quantify the roles of evaporation, surface interception, and deep infiltration of water in the regional hydrological cycle.

• Investigation of the utility of satellite data inversion algorithms in the particular context of this region. It will be necessary to obtain in situ observations to calibrate these algorithms. Thereafter, spatial extrapolation techniques must be developed.

A final example of these large-scale studies is the International Global Atmospheric Chemistry Program field projects involving rates of exchange of trace gases between representative tropical biological environments and the atmosphere. In addition, this series of experiments will assess the impact of land use changes such as cropland expansion and forest harvesting on the rates of emission. One of these experiments, the Atmospheric Boundary Layer Experiment (ABLE), consists of expeditions that seek to study the rate of exchange of materials between the earth and its atmospheric boundary layer and the processes by which gases and aerosols are moved between the boundary layer and the free troposphere. These expeditions are conducted in ecosystems of the world that are known to exert a powerful influence on global atmospheric chemistry and that, in some cases, are undergoing profound changes as a consequence of natural processes or human impacts. Several such experiments are planned in the next decade.

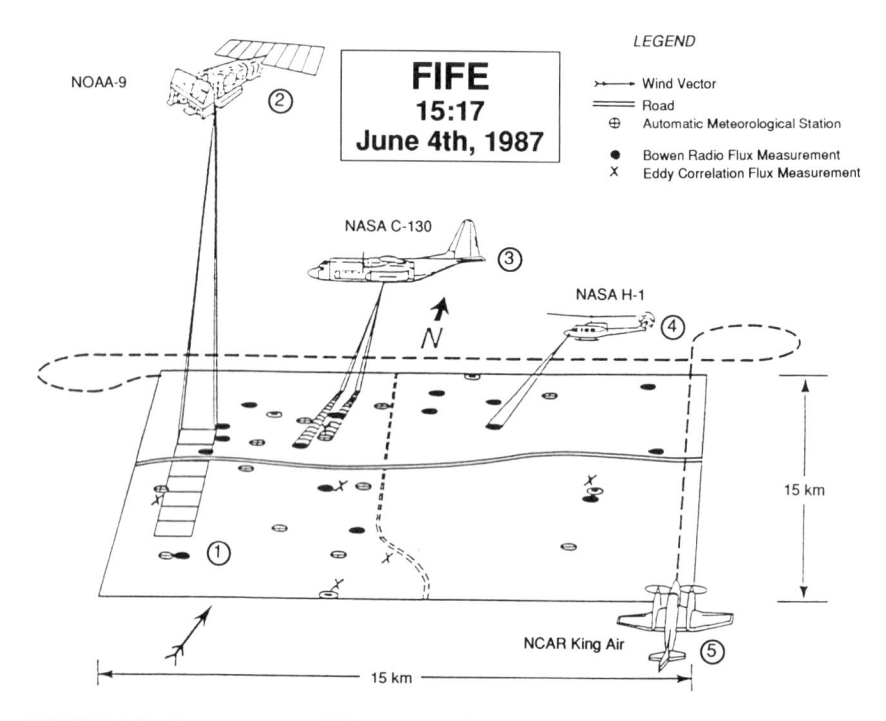

FIGURE 8.2 Situation at the FIFE site at 1517, June 4, 1987; time of the NOAA-9. (1) Surface flux stations and automatic meteorological stations monitor surface fluxes and near-surface meteorological conditions. (2) NOAA-9 satellite scans the site at 1-km resolution. (3) NASA C-130 traverses the site at 5000-m above ground level, taking scanner and sun photometer data. (4) NASA helicopter hovers above preselected site at 250-m above ground level and acquires radiometric data. (5) NCAR King Air collects eddy correlation data at 160-m above ground level. (Source: NASA, 1988.)

Surface Observation Networks

Several surface observation networks have been established around the globe to monitor the baseline characteristics of the surface and the atmosphere. In this regard, the networks organized by the World Meteorological Organization (WMO) and the U.N. Environment Programme (UNEP) are noteworthy. At the same time, the WCRP has initiated studies to validate and assess the accuracy of the data base, and the Global Monitoring for Climate Change (GMCC) of NOAA produces occasional updates of the global climate trends based on data from these networks.

Noteworthy for the USGCRP studies is the Background Air Pollution Monitoring Network (BAPMON), which was established in 1970 as one of the WMO's early activities in the field of air pollution. It has since become an important component of the Global Environment Monitoring System (GEMS).

The function of BAPMON as outlined by the WMO Executive Council Panel of Experts in 1982 is "to obtain measurements on a global and regional basis of background concentrations of atmospheric constituents which may affect environmental pollution or climate." From the variability in time and space and the possible long-term changes reflected in these data, it will be possible to assess the influence of human and natural occurrences on the composition of the atmosphere. Such information is required

• to study the effects of atmospheric composition on climate, and to predict future climatic change due to future human activities;
• to aid in the study of the mechanisms of long-range atmospheric transport and deposition of potentially harmful substances; and
• to aid in the study of the biogeochemical cycles of important constituents in order to establish a sound basis for assessing human impacts on these cycles and for making predictions of possible impacts on the environment.

At the end of 1990, some 94 countries were participating in the BAPMON program, with 216 stations either providing data (166), in preparation (12), or in planning (38). Stations are categorized as global (16), continental (10), or regional (190).

Global background air pollution stations document long-term changes in atmospheric composition likely to affect the weather and the climate. These stations are located in areas where no changes in land use are anticipated for at least 50 years within 100 km in all directions from each station.

Although the concept of BAPMON is basically sound, measurements made by different groups within the network need to be intercompared and standardized on an ongoing basis. All relevant data and information on procedures should be deposited and maintained in a permanent archive that is accessible. Careful attention needs to be paid to the validation of the atmospheric transport model, particularly in relation to vertical mixing and interhemispheric exchange. Inert tracers of known source strength such as chlorofluorocarbons, and existing satellite measurements of atmospheric water vapor and tropical winds may be useful here. Isotopes should be measured throughout the network as soon as competent staff can be trained and adequate facilities made available. The vertical profile of the seasonal cycle and year-to-year variations of carbon dioxide should be measured at more latitudes.

Networks also exist to monitor and conduct process studies on ecological characteristics of sites around the world, e.g., the biosphere reserves under UNESCO's Man and the Biosphere program. These and other sites, for example, sites within the United States such as the Long-Term Ecological Research (LTER) sites funded by the National Science Foundation and other ecological monitoring and research sites funded by other agencies, could provide ecological information relevant to global change (LTER Network

Office, 1989). In addition, the IGBP Global Change Regional Research Centers will coordinate existing networks and develop new ones in less developed countries (IGBP, 1990). The committee recommends that the plans for the establishment of IGBP Regional Research Centers be implemented as soon as possible.

International Coordination

The international community is focusing increased attention on climate and global change research, potential impacts, and response strategies. Cooperation among agencies engaged in space-based, global earth observation programs is already extensive and is pursued not only through bilateral collaboration but also multilaterally by international satellite coordination groups. Such groups coordinate multilateral missions (including the payloads of the U.S., European, and Japanese polar platforms), promote compatibility among observation systems, facilitate data exchange, and set data product standards—all of which benefit the global change user community.

One such group, the Committee on Earth Observations Satellites (CEOS), created as a result of the 1982 Group of Seven Economic Summit, is the appropriate focal point for international coordination of the space segment of global change earth observations. Its members are those government agencies with funding and program responsibilities for satellite observations and data management. Current members are NASA, NOAA, ESA, the European Meteorological Satellite Organization (EUMETSAT), and counterpart space and earth observation agencies in Japan, Canada, France, the U.K., Germany, Italy, India, Brazil, and Australia.

NASA and NOAA are proposing changes intended to strengthen CEOS interaction with both international scientific programs (ICSU's IGBP and the WCRP) and intergovernmental user organizations (the Intergovernmental Panel on Climate Change (IPCC), WMO, UNEP, and the Intergovernmental Oceanographic Commission (IOC)) with the specific goal of focusing the earth observation mission planning in space agencies on global change requirements. Scientific and intergovernmental agency representatives would be invited to participate in CEOS policy deliberations and technical coordination activities. The U.S. agencies are further proposing revitalization of the CEOS Sensor Calibration and Performance Validation Working Group to undertake important global change calibration activities, as well as the possible chartering of a Working Group on Space Networks. The CEOS Working Group on Data, chaired by NOAA, already plays an active role in standardizing data formats worldwide, achieving an international interoperable catalog system, and identifying data sets to test a proposed international network for electronic data transmission.

INFORMATION AND DATA MANAGEMENT

A data and information system for global change research must foster the process to identify global change and to evaluate its impact on human activities in order to set a course of action to mitigate harmful effects. Consequently, the USGCRP will make unprecedented demands for the assembly and dissemination of large volumes of diverse and interdisciplinary data and information.

Data System Requirements

Data system requirements for global change fall into five categories of activity:

1. Collection, processing, and analysis of past and existing global measurements to
   - establish the relative mean state of the earth system,
   - obtain measures of system variability,
   - detect change, and
   - understand large-scale interactive processes.

2. Sustained (future) global measurements to monitor and document change and to supply essential state variables for earth system models.

3. Process studies to understand particular phenomena and relate smaller-scale processes with the large-scale variables of the global system.

4. Analysis of past records, both instrumental and proxy, to
   - obtain long-term reference states of the global system,
   - document and understand secular fluctuations in the earth system and their relationship to shorter-time-scale processes, and
   - provide a basis for testing models.

5. Production of data fields (with known or uniform requirements) from assimilation and/or simulation models.

Measurements acquired on regional and worldwide scales must be merged with other, often dissimilar, data to produce analyses and products. Subsets of these data must be quality assured, documented, distributed, and archived. Contemporary and future researchers must be able to acquire and use these data in their analyses of global change phenomena.

The efficient acquisition, quality assurance, documentation, distribution, and preservation of relevant data sets of all types is crucial to the success of the USGCRP.

The section "Information and Data Management" was prepared following the outline of recommendations made by the NRC Committee on Geophysical Research (1990).

Kinds of Data Needed

Data needed for broad global change research run the gamut from site-specific to truly global data sets, as described earlier in this chapter. Disciplines involved include the geosciences, ecology, biology, and socioeconomics. Many data sets will cross disciplines, and people of diverse talents and skills will be required to assemble them. Time series data of all types are prominent in global change research to detect past and present trends. Proxy data are necessary when direct measurements are impossible. The need for accuracy is important, as some of the predicted changes will be small and take place over long periods of time. Often data sets will be enormous, owing to the resolution and spatial scale needed to address global issues.

Functions of a Data and Information System

The prime function of a data and information management system is the stewardship of the data and information, with all its ramifications. Such information is costly to acquire. Its safekeeping must not be left to chance.

Other functions may vary depending on programmatic goals, attributes of the thematic data or programmatic issues, and researcher needs. These functions include, but may not be limited to, the following:

- Data preservation to ensure the long-term stewardship of research data.
- Data distribution. Data must be easily accessible by the world research community with as few restrictions (including cost) as possible.
- Data integration and product production. The system must be able to integrate data within and across disciplines to create data products for use by the research community and policymakers.
- Data quality assurance using high standards to maximize the application of data to answer global change questions.
- Provision of data documentation. Data sets must be fully documented (documentation is often termed "metadata") to ensure their complete understanding today and their usability in the future.
- Data identification and acquisition. The system must take an active role with scientists to identify data sets useful for global change research.
- Provision of a programmatic focus for data management in order to focus the flow of information necessary to conduct global change research.
- Selective data retrieval. It must be possible to retrieve selectively data relevant to a user's needs.
- Standardization of procedures to ensure that standards for quality assurance, documentation, and distribution are similar among system components.

Many requests to the data management system will be for derived products such as analyses or edited data collections in association with descriptive text or graphical material, rather than raw observational data. Thus the system must provide information as well as data. It must keep the raw data as well, to provide the material for future reanalyses. It must play an active role in the generation, acquisition, quality control, dissemination, and retention of value-added products.

Creating a New System

A new system for data and information management should begin with the establishment of objectives. What data sets are needed to describe global change? What are the highest priorities? Setting those objectives and priorities is an activity that goes beyond data management. The fundamental scientific design of the USGCRP should include setting objectives for the data and information that will be needed. These objectives must be set with a data management input into the scientific process.

A network of discipline-oriented data centers is an important and necessary component of the system to support global change research. Data centers should be created for those disciplines important to global change research that are not included in the network of national data centers. The network can be augmented by establishing new centers or by expanding the purview and resources of existing centers.

The extraordinary information requirements that the global change program will make on existing data management elements will necessitate augmenting the existing system with a new mechanism to handle these data. This mechanism is a data management infrastructure. One does exist today, but it is incomplete, inadequate, and poorly supported. It must be strengthened. If the USGCRP is to be a success, a strong data management system must exist to support it.

The new system should build upon itself. As a first step, we must make sure that the existing components work. Making them work is not a technical challenge but one of will and resources. Existing U.S. environmental data management units must be improved, restructured, or replaced. Then we can move on to create the more complex system. New components must be created. Some existing institutions can serve as models. Above all, data and information management must be adequately supported.

Agencies must shoulder the responsibility of providing for the stewardship of the data they generate. Data management should be considered at the outset of every project, explicitly defined, and adequately budgeted for the life of the project. Arrangements should be made for the long-term archiving of the data.

A new system should demonstrate success through practical prototypes. Confidence will be built by proving accuracy, by showing that things can be done right, and by producing some early products of value. This can be done by beginning feasible pilot projects of great importance.

The Master Directory project sponsored by the Interagency Working Group on Data Management for Global Change is an excellent example. This project is creating a high-level directory of data sets related to global change held by many agencies and institutions. Its success depends on common information standards among centers and on operational computer network links.

Scientific Involvement

If we look at existing data and information management activities, we find that successful data systems and centers often combine data management with scientific use. Successful data centers not only work with scientists but also have their active support. Users support the development of the data system and provide feedback. A successful system must involve the scientific user community at all stages of development and operation. Without that support and involvement, the data system is unlikely to meet the needs of the program (Data Management and Computation Committee, 1982).

Achieving effective scientific involvement will not be easy. Unfortunately, much of the scientific community is not aware of the need to be involved in any unified global change data and information management approach. This attitude is part of a pattern: data management has long been considered a secondary aspect of research. Since data will be such a critical element in global change research, a change of attitude is essential. Fortunately, there are many signs that this change is taking place, both in research scientists and in sponsoring agencies. For example, an elaborate system is being devised to handle the data that will be generated by NASA's EOS program. The EOS Data and Information System (EOSDIS) is being planned in parallel with the EOS program in an attempt to ensure that the scientists will be involved in the storage and archiving of data as well as the analysis.

Active researchers must be participants in the process. They should define needs and create the framework for a data and information system to meet those needs. They should help establish procedures and data centers. It will not be enough for them simply to assent to what a group of data technologists are creating.

There must be incentives for researchers to be involved. The system must respond to scientists' needs. It must be perceived as the optimal way to do research with data. A simple feedback process will have a beneficial effect. When advice is sought and listened to, there is an incentive for involvement.

Not only should incentives be created, but existing disincentives should be removed. User fees above a minimal cost for reproduction for scientific use of data constitute an existing disincentive. The nature of global problems requires access to large data sets. If their costs make this prohibitive, then exploratory research will be obstructed. For example, Landsat data are currently so expensive that the data are generally beyond the reach of the research community. The global change data system should have the lowest possible user fee structure. Data should be free wherever possible.

Data Directories

Many scientists face major obstacles in finding out what pertinent data are available; in some situations, obtaining the data may be a practical impossibility. In other cases, inadequate documentation makes personal contact with the primary user imperative.

There is a need for all centers holding data acquired through federal funding to provide well-documented information about the extent of their holdings and the accessibility of the data. Furthermore, data interchange among agencies will become a major issue as global change programs require data from ever-wider sources. The lack of interoperability among data directories is a serious deficiency of the current system. It must be addressed if the USGCRP is to draw upon existing and future data holdings.

Locating data, both nationally and internationally, will be helped by the establishment of a centralized data directory. This directory will have information about information. It will be created by a joint effort of the national data centers and, eventually, international data sources. The data directory should be electronically accessible and user friendly. It should provide as much information about data sets as possible, including location, access policies and procedures, and information about the data's completeness, accuracy, general usefulness, documentation, and limitations. It should be free to use.

Data Submission

There are valuable data sets that remain in the custody of individual research groups or even individual investigators, with the consequent acceptance by them of the data-handling task. Some of these data sets are widely known and accessible; others are not. There are many reasons for the failure of individual scientists to provide data to the data centers. Among these are a desire for exclusive access to data, the reluctance to divert time and effort away from research in order to clean up a messy research data set, and lack of awareness of the existence of appropriate repositories or of the importance of depositing data in such centers.

In general, the research community is not sufficiently aware of the importance of ensuring the availability of environmental data. Until this changes, potentially valuable data sets will continue to be lost as primary users retire, relocate, or move on to other projects. There also remains the question of data ownership: how long should recently collected data remain under the exclusive control of the scientist(s) who collected the data? This issue is being addressed by the federal agencies involved in the global change program.

Quality Assurance and Documentation

Data are frequently separated from information about the data. This is an unfortunate by-product of the explosion over recent decades of digital data and techniques for handling them. Information about the algorithms used for a derived product, quality control procedures, comparisons with independent measurements, reviews by outside experts, and so on, permit the user to judge the reliability or value of the product for a particular application and therefore should be an inseparable part of the data. The same is true for original data in terms of calibration, quality control flags, station histories, and so on.

The quality assurance and documentation standards of data sets important to global change research must be upgraded. Quality assurance and documentation should be at the heart of a data management system supporting the global change program. Only after extensive testing by independent reviewers should important global research data sets be considered accurate. Depending on the data involved, the effort can be extensive and often can be the single most expensive step in processing a data set for distribution. This process, analogous to the independent peer review of a journal article, maximizes the integrity of the information. It is necessary for USGCRP. Future research and policy decisions will rest in part on these important data sets.

Documentation must do more than describe the values represented in each field and the format information to read the data tape. It must fully document the data set from all possible points of view. Data documentation must pass the "20-year test." That is, 20 years from now will someone not familiar with the data or how they were obtained be able to fully understand and use the data solely with the aid of the documentation archived with the data set? This is a tough test, and yet one that must be passed for many of the data collections if long-term global environmental programs are to be successful.

Cooperation and Sharing

For most global change studies, regional and global data and information will be required. No one nation, agency, or institution will be able to gather the appropriate data without cooperation from other nations, agencies, and institutions.

In the United States, the USGCRP will depend on scientists sharing their data with each other. The timely submission of data to national centers requires a policy to ensure it. The policy must recognize the needs of principal investigators to protect their intellectual investment and must encourage their continued efforts to collect useful global change data.

REFERENCES

Data Management and Computation Committee, National Research Council. 1982. Data Management and Computation, Vol. 1: Issues and Recommendations, National Academy Press, Washington, D.C.

Earth System Sciences Committee. 1988. Earth Systems Science: A Closer View. National Aeronautics and Space Administration, Washington, D.C.

Geophysical Data Committee, National Research Council. 1990. A U.S. Strategy for Global Change Data and Information Management. In review.

Interagency Working Group on Data Management for Global Change. 1990. Recommendations from an Interdisciplinary Forum on Data Management for Global Change. Office for Interdisciplinary Earth Studies, UCAR, Boulder, Colo.

International Geosphere-Biosphere Program (IGBP). 1990. The Initial Core Projects. Report to the Second Advisory Council for the IGBP. In preparation.

Long-Term Ecological Research (LTER) Network Office. 1989. 1990's Global Change Action Plan: Utilizing a Network of Ecological Research Sites. LTER Network Office, Seattle, Wash.

Morel, P. 1990. Satellite Observations and Global Climate Change Research. Space and Global Environment, Paris.

National Aeronautics and Space Administration. 1988. NASA Earth Science and Applications Division: Program Plans for 1988-89-90. NASA, Washington, D.C. 133 pp.

Panel to Review NASA's Earth Observing System in the Context of the U.S. Global Change Research Program. 1990. Preliminary Report. NRC, March 30.

Potential of Remote Sensing for the Study of Global Change. 1987. S.I. Rasool (ed.), Advances in Space Research, Vol. 7, No. 1, Pergamon Press.

Rasool, I., and D. Ojima (eds.). 1989. Pilot Studies for Remote Sensing and Data Management, IGBP Report No. 8, Stockholm, Sweden.

APPENDIXES

Appendix A
List of Participants in the Workshop on Human Interactions with Global Change

Richard Berk
Department of Sociology
University of California
405 Hilgard Avenue
Los Angeles, CA 90024-1551

William Clark
Science, Technology and Public
 Policy Program
Kennedy School of Government
Harvard University
79 Kennedy Street
Cambridge, MA 02138

Pierre Crosson
Resources for the Future
1616 P Street, N.W.
Washington, DC 20036

Faye Duchin
Director of the Institute for
 Economic Analysis
269 Mercer Street
New York, NY 10003

Jae Edmonds
Battelle Pacific Northwest Laboratory
901 D Street, S.W., Suite 900
Washington, DC 20024

Robert C. Harriss
Institute for the Study of Earth,
 Ocean and Space
Science and Engineering Research
 Building
University of New Hampshire
Durham, NH 03824

Susana Hecht
GSAUP
University of California, Los Angeles
Los Angeles, CA 90024

Robert Kates
World Hunger Program
Brown University
Box 1831
Providence, RI 02912

Ronald Lee
Graduate Group of Demography
University of California
2232 Piedmont Avenue
Berkeley, CA 94720

Roberta Balstad Miller
National Science Foundation (SES)
1800 G Street, N.W., Room 316
Washington, DC 20550

John D. Montgomery
Kennedy School of Government
Harvard University
79 Kennedy Street
Cambridge, MA 02138

Harold Mooney
Department of Biological Sciences
Stanford University
Stanford, CA 94305-2493

Granger Morgan
Department of Engineering and
   Public Policy
Carnegie-Mellon University
Schenley Park
Pittsburgh, PA 15213

Vicki Norberg-Bohm
Kennedy School of Government
Harvard University
79 Kennedy Street
Cambridge, MA 02138

Peter J. Parks
Center for Resource and
   Environment Policy Research
214 Biological Sciences Building
Duke University
Durham, NC 27706

Stephen J. Pyne
15221 N. 61st Avenue
Glendale, AZ 85306

Steve Rayner
Global Environmental Studies
   Center
Oak Ridge National Laboratory
Oak Ridge, TN 37381-6206

John Richards
Department of History
Duke University
Durham, NC 27701

Jennifer M. Robinson
Department of Geography
Pennsylvania State University
302 Walker Building
University Park, PA 16802

Richard Rockwell
Social Science Research Council
605 Third Avenue
New York, NY 10158

V.W. Ruttan
Department of Agriculture and
   Applied Economics
231 Classroom Office Building
University of Minnesota
1994 Buford Avenue
Sanford, MN 55108

Sam H. Schurr
Electric Power Research Institute
3412 Hillview Avenue
Palo Alto, CA 94303

Robert Stavins
Kennedy School of Government
Harvard University
79 Kennedy Street
Cambridge, MA 02138

B.L. Turner II
Graduate School of Geography
Clark University
950 Main Street
Worcester, MA 01610

Anne Whyte
International Development Research
   Centre
P.O. Box 8500
Ottawa, Ontario, Canada K1G 3H9

Staff

Ruth DeFries
Committee on Global Change,
  HA 594
National Research Council
2101 Constitution Avenue, N.W.
Washington, DC 20418

John S. Perry
Committee on Global Change,
  HA 594
National Research Council
2101 Constitution Avenue, N.W.
Washington, DC 20418

Samuel Rod
Program Office, NAS 310
National Academy of Engineering
2101 Constitution Avenue, N.W.
Washington, DC 20418

Hedy Sladovich
Program Office, NAS 309
National Academy of Engineering
2101 Constitution Avenue, N.W.
Washington, DC 20418

Paul Stern
Study Director
Commission on Behavioral and
  Social Sciences and Education,
  HA 184
National Research Council
2101 Constitution Avenue, N.W.
Washington, DC 20418

Appendix B
A Selective Literature Review on the Human Sources of Global Environmental Change

by Vicki Norberg-Bohm

This appendix, written in preparation for the Workshop on the Human Interactions of Global Change, briefly reviews scholarship in the two areas identified by the U.S. Committee on Global Change as initial priorities for research into the human interactions with global change: land use changes and industrial metabolism. Although an enormous number of relevant studies have been done, it is not within the scope of this document to provide a thorough review of all this work. This appendix strives to be illustrative rather than exhaustive. In general, only more recent works are discussed, and the emphasis is descriptive rather than critical. The goal of this review is to provide a starting point for determining where an extension of current research directions and methods will provide usable knowledge for global change studies, and where (and what) new directions or methods are needed.

As was the case in the main body of this report, this literature review uses the categories of data, process, and synthesis as an organizing framework. Section 1, integrative modeling studies, highlights examples of research that has developed a synthetic framework or model capable of generating consistent scenarios of global environmental change. Section 2 describes studies on industrial transformations of material and energy, while section 3 describes studies on land use transformations. The main focus in these two sections is on process studies. Section 4 provides a discussion of several data bases that have been developed for global change studies. Finally, because many of the models depend on population estimates, section 5 provides a review of global population models.

This review is organized around illustrative major studies, or groups of studies of a similar nature. Because most studies do not singularly contribute only to data, process, or synthesis, each study (or group of studies) is reviewed for the contribution it makes in each of these three general areas.

In some cases, the category "synthesis" is not included because the study being reviewed makes no major contribution to synthesis, as the term has been narrowly defined for the purposes of this document.

1 INTEGRATIVE MODELING

The studies highlighted in this section are the most ambitious examples of research that has developed a synthetic framework or model capable of generating scenarios of the human activities driving global environmental change. The output from these models describes the changes in emissions or in physical and biological variables (i.e., environmental transformations) caused by alternative development paths.

1.1 *Impacts of World Development on Selected Characteristics of the Atmosphere: An Integrative Approach* (Darmstadter et al., 1987)

This study focuses on "development, atmospheric emissions associated with development, and atmospheric impacts caused by emissions." It was an interdisciplinary collaborative effort that grew out of discussions at a conference sponsored by the Sustainable Development of the Biosphere Program at the International Institute for Applied System Analysis (IIASA). In addition to its synthetic approach, its major contribution is further development and implementation of a qualitative methodology for ranking the relative contribution from various sources, and for historical assessments of fluxes of key chemicals. The categories of atmospheric impact that it examines are photochemical smog, precipitation acidity, atmospheric corrosion, and stratospheric ozone depletion.

Data. The study constructed a data base of emissions of CH_4, NO_x, SO_x, HCl, and sea salt on a regional basis, and of CH_4, CO, NO_x, N_2O, and CFCs on a global basis. Data is for the years 1800 to 1980, every 30 years, excluding 1830. In some instances, no data was available for the nineteenth century. Sources of emissions are metallurgical and certain other industrial operations, coal production and use, petroleum production and use, biomass combustion, and emissions from vegetation and soils. Regions are the Northeastern United States, Europe, the Gangetic Plain of India, and the Amazonian basin of Brazil.

The study contributes new data in historical estimates of land in wet rice cultivation and historical emissions from combustion; the flaring of natural gas; and smelters, cokers, and other industrial processes.

Process. This study performs a historical reconstruction of industrial practices and technologies to determine emissions from industry and energy sources. See section on "Industrial Metabolism" in chapter 4 for a description of this materials balance approach. Future demand is based on IIASA "conventional wisdom" reference scenarios (Anderberg, 1989). Two scenarios were examined: one assumed constant emission coefficients, the other a 1 percent yearly rate of decline in emission coefficients (i.e., no technological change and constant rates of change).

Synthesis. This study is synthetic in three respects: (1) It combines an analysis of historical emission coefficients with data on levels of human activity to develop a historical data base of emissions. (2) It combines information on emission factors with scenarios of future development to develop qualitative assessments of future environmental flows and thus the degree of environmental degradation. (3) It includes emissions from both industry and land use in its analysis.

1.2 *Long-Term Global Energy and CO_2 Model* (Edmonds and Reilly, 1983)

A model, the Institute for Energy Analysis of Oak Ridge Associated Universities model, was developed at Oak Ridge Associated Universities for the U.S. Department of Energy to examine future scenarios of CO_2 emissions. "The long-term, global energy-CO_2 model was developed to provide a consistent and conditional representation of economic, demographic and energy interactions (Edmonds and Reilly, 1983)."

Although this model looks only at emissions from fossil fuel use, it is a prime example of synthesis in that it combines energy supply and demand scenarios (driving forces include economic and demographic factors) with CO_2 emission factors derived from an understanding of various combustion processes to produce estimates of CO_2 emissions. The end product is a model that can be used to analyze future scenarios of CO_2 emissions.

This model has been used extensively in the analysis of future CO_2 emissions. Edmonds and Reilly (1983) and Mintzer (1987) used this model for evaluating global emissions for various types of policy intervention. Chandler (1988) used the model for evaluating policy options for reducing CO_2 emissions and achieving economic development goals for China. The authors of an EPA study, *Policy Options for Stabilizing Global Climate* (Lashoff and Tirpak, 1989), used this model as a starting point from which they made significant modifications.

A discussion of the strengths and weaknesses of using this model is found in an interchange between Keepin (1988) and Edmonds (1988).

Data. This study uses emission coefficients and elasticities calculated elsewhere.

APPENDIX B   249

Process. This is a partial equilibrium model. Critical demand assumptions include population, labor productivity, gross national product growth rates, an energy technology parameter that specifies the rate of change in energy productivity, and price and income elasticities. Critical supply assumptions include resource constraints and breakthrough costs of new technologies.

Demand in the Organization for Economic Cooperation and Development is disaggregated into three economic sectors: residential/commercial, transport, and industrial. All other regions are modeled as a single sector. The model makes projections to the year 2100 in 25-year intervals. The globe is divided into nine regions. The model includes six primary fuel sources and four secondary fuel sources, as well as biomass, shale oil, and synfuels. The outputs of the model include primary and secondary fuel mixes; a variety of trade, price, and development indicators; and CO_2 emissions.

Synthesis. This model combines energy supply and demand (driving forces include economic and demographic factors) with CO_2 emission factors derived from an understanding of various combustion processes to produce estimates of CO_2 emissions. The model is constructed to facilitate the examination of alternative future energy paths based on different assumptions about prices, population, economic growth, technological change, and supply constraints.

1.3 *Policy Options for Stabilizing Global Climate*, U.S. Environmental Protection Agency (Lashof and Tirpak, 1989)

This report was written in response to a congressional request to examine "policy options that if implemented would stabilize current levels of atmospheric greenhouse gas concentrations." One of the major goals of the study was "to develop an integrated analytical framework to study how different assumptions about the global economy and the climate system could influence future greenhouse gas concentrations and global temperatures."

Data. This study has compiled the best available estimates of current emissions of all greenhouse gases. In a few cases, new data bases were developed, such as an energy end use data base (Mintzer, 1988, for industrialized countries; Sathaye et al., 1988, for developing countries).

Process. This study has compiled the best available estimates of emission coefficients for all greenhouse gases. Future activity levels are determined by population growth, economic development, and technological change.

The study develops four scenarios of the future. These are based on two different patterns of economic development and technological change, each examined with and without policy intervention to stabilize climate change.

A sensitivity analysis is performed on values for the key variables in these scenarios.

Assumptions regarding population growth rates, economic growth rates, and oil prices are developed as follows: Population estimates were developed from Zachariah and Vu (1988) of the World Bank and from the U.S. Bureau of the Census (1987). The primary source for economic growth rates was the World Bank (1987). Oil prices were taken from the U.S. DOE (1988).

Synthesis. The study combines models of activity levels with information on emission coefficients to develop an analytical framework that relates the underlying forces of population, economic development, and technological change to the emissions of all the important greenhouse gases. The study uses several detailed models of individual components to inform this general framework.

Four modules are used to calculate emissions. Attention was paid to developing consistent scenarios, but there are no explicit feedbacks between modules. The four modules are briefly described below. A more detailed description can be found in the appendix of the EPA report (Lashof and Tirpak, 1989).

1. The energy module is based on a considerably modified Edmonds-Reilly model (developed by ICF) and two end use studies (Mintzer, 1988, for industrialized countries; Sathaye et al., 1988, for developing countries). The end use studies are used to project demand in the year 2025. This estimated demand in turn is used to anchor the demand estimates that are calculated for other years using the modified Edmonds-Reilly model.

2. The industry module is based largely on the EPA's CFC model (U.S. EPA, 1988a). Non-CFC industry emissions (from landfills and cement manufacture) are calculated as simple estimates of population and per capita income.

3. The agriculture module uses the IIASA/IOWA Basic Linked System to calculate agricultural production and fertilizer use. This model was first developed at IIASA's Food and Agriculture Program. It was modified by the Center for Agriculture and Rural Development at Iowa State University to extend the time horizon to the year 2050 (Frohberg and Van de Kamp, 1988; Fisher et al., 1988). Emission coefficients are derived from the literature.

4. The land use and natural source module uses the terrestrial carbon model developed at the Woods Hole Marine Biological Laboratory to calculate CO_2 emission factors for land use changes. CO and N_2O emissions are scaled based on CO_2 emissions. Natural emissions (from forest fires, wetlands, soils, oceans, and fresh water) are based on values from the literature and generally held constant.

APPENDIX B

The study uses two concentration modules to calculate atmospheric concentrations and temperature increases based on scenarios of emissions.

1.4 *Future Environments for Europe: Some Implications of Alternative Development Paths* (Stigliani et al., 1989a,b)

This study is a regional case study sponsored by the Sustainable Development of the Biosphere Program at IIASA. "The purpose of this study is to provide new insights into the long-term management of the European environment during an era of fundamental transitions in technologies, climate, and scale of effects." The specific objectives of the study include developing a method for examining regional environmental problems 40 years into the future, learning about the major environmental problems that would be facing Europe in this time frame, and developing tools to improve the management of the environment in the long term. The study considers land use transformations and industry and energy transformations in its assessment.

Data. The study uses current data (1980) on activity levels for population, energy, industry and transportation, agriculture, and forestry. These provide the starting point for scenario development.

Process. The study constructs several socio-economic development paths (scenarios of the future) for Europe. These paths describe future trends in population, energy, industry and transportation, agriculture, and forestry. These trends in turn cause changes in the environment. The environmental components analyzed include climate, hydrology, atmospheric pollution and regional acidification, soil quality, water quality, biota, and land use.

This study develops a scenario based on conventional wisdom and several based on not impossible alternatives to the most likely scenario. These alternatives are based on surprises, or turning points from the conventional wisdom scenario. The study uses a qualitative framework similar to that used in the Darmstadter et al. (1987) study for presenting the seriousness of the environmental consequences of four different development paths.

Synthesis. The socio-economic scenarios are used to drive development. The study describes changes in the environment based on these scenarios.

1.5 *Project Proposal: Strategies for Environmentally Sound Development: An Input-Output Analysis* (Duchin, 1989c)

This describes a project that was recently begun at the Institute for Economic Analysis at New York University. "The objective of the proposed study is to identify and evaluate concrete, consistent, economically feasible strategies for environmentally sound development, that is, to examine alternative approaches to reducing poverty over the next 50 years while also reducing global pollution." The resulting analysis will be based on a world input-output model.

The analysis will incorporate detailed technical process information and provide quantities and geographic distribution of pollutant emissions under various scenarios as one of its outputs.

2 INDUSTRIAL METABOLISM: TRANSFORMATION OF MATERIALS AND ENERGY

Industrial metabolism can be defined as the production and consumption processes of industrial society. These processes include extraction, processing, refining, use, and dispersion of fossil fuels and minerals. These processes transform materials and energy into emissions to the environment and are thus a major source of global environmental change in industrialized societies. One of the goals of this report is to define research initiatives that will improve understanding of how the historical and current industrial metabolism have caused and are causing environmental change. Equally important is gaining an understanding of the dynamics of industrial metabolisms: what are the factors causing changes, how have they changed over time, and what are possible future industrial metabolisms.

This section is divided into four subsections: materials balance studies, trends in material and energy intensity, long wave studies, and global energy modeling.

2.1 Materials Balance Studies

The materials balance approach is based on the concept of conservation of mass (i.e., the first law of thermodynamics). It tracks the use of materials and energy from "cradle to grave." In other words, it follows them from extraction through various transformation processes to disposal and their final environmental destination. It is a tool that allows economic data to be used in conjunction with technical information on industrial processes to describe chemical flows to the environment. For a discussion of this methodology see Ayres (1989) and Ayres et al. (1989).

Some important conclusions that have been drawn from applying this type of analysis are as follows: (1) Major sources of environmental pollutants have been shifting from production to consumption processes. (2) Large numbers of materials uses are inherently dissipative, spreading widely in the environment.

2.1.1 The Hudson-Raritan Study (Ayres et al., 1988; Ayres and Rod, 1986)

This is the most far reaching study of this type. It provides a historical reconstruction of major pollutant levels in the Hudson-Raritan Basin from 1880-1980. The methodology was a materials balance approach. The major contribution of this work is in the framework it provides for developing data of pollutant loadings using process information and economic data.

Data. This study provides current and historical (1880-1980) pollutant loading data for the Hudson-Raritan river basin for heavy metals (silver, arsenic, cadmium, chromium, copper, mercury, lead, and zinc), petroleum and coal, and for chemicals and other wastes (chlorinated pesticides, chlorinated herbicides, chlorinated phenols, polynuclear aromatic hydrocarbons, oil and grease, carbon, nitrogen, and phosphorus).

Process. This study developed process-product flows for heavy metals that describe the location and form from extraction through consumer end use to the disposal of these materials. It used historical data of how processes changed over time to determine the level of different types of production activities. It used emission coefficients from the literature on production emissions. There is little information in the literature on consumption emissions; thus the study used an ad hoc choice of consumption emission coefficients. The runoff estimation model is a modified version of that developed by Heany (Heany et al., 1976).

Synthesis. This study implements the materials balance framework for one region. It serves as an example of how data on pollutant loadings can be developed using process information and economic data.

2.1.2 Other Studies

Several other studies have examined the processes of transformation of materials and energy and developed data on emissions.

1. *Impacts of World Development on Selected Characteristics of the Atmosphere: An Integrative Approach* (Darmstadter et al., 1987). This study provides a historical reconstruction of emissions of CO, SO_x, N_2O and NO_x, and CH_4 for the years 1880 to 1980 for four regions. For a more detailed discussion of this study, see section on "Materials Balance Studies."

2. "Carbon Dioxide from Fossil Fuel Combustion: Trends, Resources, and Technological Implications" (Rotty and Masters, 1985). This study develops global emissions of CO_2 from fossil fuel combustion for the years 1860 to 1982.

3. *The Study of Chemical Pollution and Its Sources in Dutch Estuaries and Coastal Regions, a Proposal for a Collaborative Agreement* (Shaw, 1989) will be using a materials balance framework. It is just beginning as a collaborative project between The Netherlands' Ministry of Public Housing, Physical Planning and Environment, The Netherlands' National Institute of Public Health, and the International Institute for Applied Systems Analysis. An interesting feature of this study is the use of the RAINS model (developed at IIASA to trace regional pollution for acid rain) to determine heavy metal loadings from atmospheric releases.

2.2 Trends in Material Intensity and Energy Intensity

Material intensity is defined as the mass of a material per unit of GNP or per capita. Similarly, energy intensity is defined as the energy per unit of GNP or per capita. Energy intensity is also defined as the primary energy per unit of useful energy or end use service. In sum, material intensity and energy intensity are defined as the quantity of material or energy consumed per unit of value created. Trends in material intensity and energy intensity are determined by changes in the amount and types of goods and services that are produced and consumed, the efficiency of energy and material use in the production and consumption process, and the substitution of materials within the same good (e.g., plastic instead of steel in automobiles). In other words, these trends are determined by the structure of the economy, the income level, and technology. The topic of whether the industrialized countries are experiencing a decline in material intensity and energy intensity, a trend called "dematerialization," is relevant to scenarios of future environmental effects from industrialization.

This section reviews studies of material intensity and energy intensity and studies of substitution of one material for another.

2.2.1 *Materials, Affluence, and Industrial Energy Use* (Williams et al., 1987)

This study focuses on the trends in the use of materials in the United States. It concludes that there is indeed a trend toward dematerialization in the United States.

Data. This study is based on about 100 years of data on prices and consumption of steel, cement, paper, ammonia, chlorine, aluminum, and ethylene in the United States, in units of kilograms, as well as on data for low- and intermediate-volume metals, including copper, lead, zinc, manganese, chromium, nickel, tin, molybdenum, titanium, and tungsten.

Process. This study concludes that "the United States is passing the era of materials-intensive production and beginning a new era of economic growth dominated by high-technology products having low materials content." Dematerialization is the result of a structural shift in the United States, which is based on the level of income. Analyses of data show that reduced energy use per unit GNP in the United States is half from structural changes and half from energy efficiency improvements. The authors postulate three stages and a bell curve for the materials use cycle. This has implications not only for materials flows, but also for energy use. The result is that industrial demand for energy may be zero growth or negative.

The maturing of basic materials use in the United States is attributed to improvements in efficiency of materials use, substitution of cheaper materials or materials with more desirable characteristics for traditional materials, saturation of bulk markets for materials, and shifts in the preferences of consumers at high income levels for goods and services that are less materials intensive. Recycling can achieve greater market share as demand growth for a material decreases.

This study examines in detail the trends in materials use for steel, ethylene and plastics, aluminum, pulp and paper, minor metals, and "new age" materials.

2.2.2 "Dematerialization" (Herman et al., 1989)

This essay examines the question of whether dematerialization is occurring, and what is a meaningful definition of dematerialization with regards to the environment. The authors suggest defining dematerialization as "the amount of waste generated per unit industrial product." Their goal is to look at forces "beyond the obviously very powerful forces of economic and population growth."

Data. The authors provide data that shows that solid waste streams from consumers have been growing.

Process. The authors identify product life as a key factor in dematerialization and identify several product traits that are important in determining product life, including quality, ease of manufacture, production cost, size and complexity of the product, ease of repair or replacement, and size of waste stream. They draw a distinction between the dematerialization of production and consumption.

2.2.3 "Energy Use, Technological Change, and Productive Efficiency: An Economic-Historical Interpretation" (Schurr, 1984)

The goal of this paper is to explain the simultaneous occurrence of rising total productivity, low energy prices, and declining intensity of energy use. This work builds upon, and updates, research originally reported in the

1960 Resources for the Future book, *Energy in the American Economy*, by the author and associates.

Data. This analysis is based on data of energy use, capital and labor inputs, and productivity for the past century.

Process. The intensity of energy use has risen in relation to labor and capital inputs, but has dropped in relationship to total output since 1920. The explanation for this apparent paradox is based on an energy-technology-productivity connection thesis. The characteristics of energy supply—low cost, abundance, and enhanced flexibility in use—sets the stage for discovery, which quickens the pace of technical advance. This is reflected in labor and multifactor productivity increases, which lead to increases in total output.

2.2.4 *Energy for a Sustainable World* (Goldemberg et al., 1987, 1988)

This work presents the findings of the End Use Global Energy Project, a study by an international team of researchers. It analyzes energy demand from an end use perspective, with a focus on energy efficiency improvements that are technically possible using commercially available or near-commercial technologies. The results of this study are presented in two forms: a report containing the major findings (Goldemberg et al., 1987) and a book presenting the models and data in greater detail (Goldemberg et al., 1988).

Data. This study presents data on trends in energy and material intensity. It includes data on energy consumption disaggregated by sector, i.e. commercial, residential, transportation, and industry. Within these sectors, there is great detail on specific end uses. The study also presents large amounts of technical information on the energy efficiency of equipment, appliances, automobiles and other modes of transportation, and industrial processes. This work includes detailed case studies of the United States, Sweden, India, and Brazil.

Process. An examination of energy use in the industrialized countries leads to the conclusion that there are structural economic shifts toward less energy-intensive activities, and that there is great potential for more efficient energy use. Future scenarios of energy use in the United States and Sweden are presented. These scenarios are based on the saturation of the most energy efficient technologies that are commercially available or near commercial.

For developing countries, they examine the energy requirements for meeting basic human needs. Again, the most efficient commercially available technologies are applied.

Synthesis. The study uses technical data on energy efficient technologies in conjunction with assumptions about population and economic growth to develop scenarios of future energy consumption. The result is a normative model which shows that human needs can be satisfied (including improved standards of living) with much lower energy consumption than projected by "conventional wisdom" scenarios.

2.2.5 *Toward a New Iron Age* (Gordon et al., 1987)

This book is about quantitative modeling of resource exhaustion. Its goal is to analyze future patterns of resource exhaustion, substitution, and associated price paths. The key contribution of this book is its integration of geology, substitution, and recycling, i.e., combining science, economics, and engineering.

Data. Copper resources of 48 continental U.S. states, by ore grade. Data on production and price of copper products and copper substitutes.

Process. The framework of analysis is based on general equilibrium principles. This framework is represented by a linear programming optimization model. The study estimates a supply function for copper based on a detailed assessment of U.S. copper deposits. The costs of alternative sources of copper and copper services relative to the cost of new copper determines the amount of substitution and recycling.

Estimates of demand. This study divided the use of copper into demand categories based on common engineering functions (ruling properties). It determined a switch price when a substitute material was less expensive. It used a logistic curve and a 30 year time for switching (based on Fisher and Pry, 1971). Two methods of cost estimation were used: expert opinion and use of product census data. The recycling module is weak, as there was little data available. For demand, they assumed a unitary income elasticity and zero price elasticity. Elasticity estimates are based on reasoning, as there were no data for empirical estimation. The study assumes GNP growth of 3 percent per annum for first 100 years, and 1 percent thereafter.

The model does not include currently unknown technologies. A sensitivity analysis found that the uncertainty about future technical advances is the most important single uncertainty in the study.

The authors conclude that in the year 2072, copper will be obtained from common rock, even after allowing for recycling. A major reservation in their study results is the assumption that the large-scale mining of low-grade resources will be acceptable (i.e., the concern is that they may have mistakenly represented environmental impacts).

Synthesis. Although this model does not extend to the environmental impacts, it is synthetic in its combination of assessments of resource availability, use of engineering and technical data, and an economic framework of demand in examining the future supply and consumption of a resource.

2.3 Studies of Long Waves (Marchetti, 1983, 1988; Marchetti and Nakicenovic, 1979; Nakicenovic, 1988)

Numerous studies have concluded that there are long-term regularities in the evolution, diffusion, and replacement of socio-technical systems. A review of these results is found in Ausubel (1989).

Data. These studies are basically empirical in nature. They have used data on technical substitution in the areas of energy (Marchetti and Nakicenovic, 1979) and transportation (Marchetti, 1983, 1988; Nakicenovic, 1988).

Process. The process model used is one of logistic substitution. In the case of two competing technologies, the historical data is fit to a logistic function to determine the characteristic time constant to go from 10 percent to 90 percent of market saturation (Fisher and Pry, 1971). For more than two competing technologies, the model is more complex, but similar in that historical data is used to determine the time constant (for a given technology to go from 10 percent to 90 percent of its eventual maximum market penetration). In terms of forecasting, the parameters derived from data on a given system are applied to forecast future behavior of that system. For a concise explanation of logistic substitution modeling, see Nakicenovic (1988).

Causal explanation is related to capital replacement, and substitution possibilities that have a total cost advantage over existing technologies, although this has not been rigorously discussed in the literature.

Theoretical economic frameworks have been specified by Peterka (1977) for centrally planned economies and Spinrad (1980) for market economies. Both models can be understood as strategic principles. For the Peterka model, the attractiveness of investment is proportional to the degree to which a technology is in use, and to a measure of economic merit. For the Spinrad model, the economic attractiveness of a technology is proportional to the inverse of the price that would have to be charged for its product.

Synthesis. Ausubel et al. (1988) used logistic substitution models to examine future emissions of CO_2. The logistic substitution model predicts that natural gas will be the dominant energy source for the next 50 years, peaking at 70 percent of world energy supply. This paper examined the emission levels based on this scenario of energy supply. It concludes that CO_2 emissions will be a problem, even in a methane economy.

2.4 Global Energy Modeling

Modeling of development-environment interactions on a global scale is probably best understood in the area of energy. Two global energy models, the IEA/ORAU model (Edmonds-Reilly) and the Nordhaus-Yohe model, have been developed specifically to explore the CO_2 emissions related to different future energy paths. Both of these models are based on a neo-classical economic framework. In addition to these two models, the IIASA energy model, which is based on an end use approach, is also discussed in this section. Global energy models are reviewed for their applicability to global change in Toth et al. (1989). A review of models of carbon dioxide emissions from fossil fuel use is found in Edmonds and Reilly (1985). Another recent review of energy models is found in Goldemberg et al. (1985).

2.4.1 The IEA/ORAU Model (Edmonds and Reilly, 1983)

This model is discussed in more detail in section 1.2.

Data. Uses emission coefficients and elasticities calculated elsewhere.

Process. This is a partial equilibrium model. Critical demand assumptions include population, labor productivity, and GNP growth rates, an energy technology parameter that specifies the rate of change in energy productivity, price and income elasticities. Critical supply assumptions include resource constraints and breakthrough costs of new technologies.

Demand in the OECD is disaggregated into three economic sectors: residential-commercial, transport, and industrial. All other regions are modeled as a single sector. The model makes projections to the year 2100 in 25-year intervals. The globe is divided into nine regions. The model includes six primary fuel sources and four secondary fuel sources, as well as biomass, shale oil, and synfuels. The outputs of the model include primary and secondary fuel mixes, a variety of trade, price and development indicators, and CO_2 emissions.

Synthesis. This model combines energy supply and demand (driving force includes economic and demographic factors) with CO_2 emission factors derived from an understanding of various combustion processes to produce estimates of CO_2 emissions. The model is constructed to facilitate the examination of alternative future energy paths based on different assumptions about prices, population, economic growth, technological change, and supply constraints.

2.4.2 Paths of Energy and Carbon Dioxide Emissions (Nordhaus and Yohe, 1983)

Data. Uses emission coefficients and elasticities calculated elsewhere.

Process. This model is based on a generalized Cobb-Douglas production function. The world is treated as one region. There are two aggregated fuel types: fossil fuels and nonfossil fuels. The model makes projections to the year 2100 in 25-year intervals. The exogenously specified input variables are population growth rate, labor productivity, rates of technological change in the energy industry, and the fossil fuel mix.

The outputs of the model are consumption and prices of fossil and nonfossil fuels, GNP, carbon emissions, and CO_2 concentration.

Synthesis. This model combines a neoclassical economic framework to determine future energy use with information on CO_2 emission coefficients to produce estimates of CO_2 emissions. This model can be used to perform a simple probabilistic scenario analysis.

2.4.3 Energy in a Finite World, IIASA (Haefele et al., 1981)

Data. Large amounts of data on energy technologies and energy resources.

Process. Energy demand is disaggregated into three sectors: industry, transport, and commercial-residential. Population and gross domestic production rates determine the activity levels in each of these sectors. Energy demand is determined based on these activity levels and a set of parameters for economic structure (industrial products), demographic structure (lifestyles), and technological structure (energy intensities). This set of parameters can be varied to examine alternative futures.

End use energy is translated into primary energy by use of a linear programming optimization model whose key variables are costs of capital, operating, maintenance, and fuels; costs, availability, and quality of resources; build-up rates; and energy production capacities.

The study looks at scenarios from 1975 to 2030, at 5-year iterations. The world is divided into seven regions. There are seven fuel types.

This is a loop of models, one for final energy demand, one for energy supply, one for impacts (economic and other) of energy use, and one for macroeconomic issues. While the models are not directly linked, they are designed to be run iteratively until a consistent scenario of supply and demand is reached. The macroeconomic model was not applied.

Synthesis. This model works toward integrating an end use model (containing microeconomic, technological, and demographic detail) with a macroeconomic

APPENDIX B

model. This model does not take the step of calculating CO_2 emissions, although they could be straightforwardly calculated from the energy projections provided by the model.

3 LAND USE TRANSFORMATIONS

Anthropogenic land transformations occur in the processes of agricultural production, mineral extraction, and human settlement. Land transformations associated with agriculture occur either through more intensive use of currently productive land, or by expanding into land that was either previously uncultivated or used for other purposes. In other words, this includes both intensive and extensive changes in land use. Land transformation processes contribute to global environmental change by affecting the flow of chemicals such as CO_2, CH_4, and N_2O, by changing physical properties such as albedo and roughness, and by changing biological properties such as biodiversity.

This section is divided into four subsections: land conversion and transformation; technical and institutional change in agriculture; regional dynamic land use models; and global agriculture and forestry production models.

3.1 Land Conversion and Transformation

This section first highlights two ambitious studies that survey land transformation; the first looking at land transformation in agriculture, the second taking a comprehensive historical look at land transformation over the past 300 years. The section then turns to studies of three land transformation processes of particular concern for global environmental change: wetland transformation, deforestation, and biomass burning. Other land transformation processes that are important to global environmental change, but that have not been reviewed below due to time constraints include desertification, irrigation, soil erosion, and urbanization. The first of these is discussed in Wolman and Fournier (1987), which is reviewed in section 3.1.1.

This section examines only direct human impacts on land. It does not look at second-order effects, where humans have caused changes in other environmental components, such as climate, which in turn cause land transformations.

3.1.1 *Land Transformation in Agriculture, SCOPE 32* (Wolman and Fournier, 1987)

This book is the result of the near-decade-long SCOPE project. The project was undertaken because of concern over the environmental effects of land transformations on land resources, and therefore on the ability to produce adequate food and fiber in the future.

Data. The first three chapters present historical data on trends in land use and agricultural production. The chapters on specific processes present data significant to the process under study. The book contains many detailed case studies. The book also contains a chapter on "Criteria for Observing and Measuring Changes Associated with Land Transformations." This discussion focuses on measures at the local level. It does not discuss what types of measures might be useful on a regional or global scale, or how to aggregate these local measures.

Process. This book begins with an overview of the types of land transformation and the ability for the land base to support population. It then has a chapter on "transformation of land in pre-industrial times" and one on "the industrial revolution and land transformation." The latter identifies the key forces causing change in agriculture over time as population growth, urbanization, industrialization, transport changes, and the role of science and the state. It then discusses three major types of farming and their evolution: Western European farming, the rice economies of Asia, and shifting cultivation and bush fallowing in the tropics.

The book contains chapters on several agricultural processes that transform the land, including wetland conversion, irrigation, mechanization, use of fertilizer, use of pesticides and insecticides, and practices that cause soil erosion. The main focus of these chapters is how these agricultural practices transform the land, and how the resulting transformations affect agricultural productivity. There is some discussion of other environmental problems related to these land transformations.

Several case studies are presented as follows:

1. Land transformation in Israel.
2. Influence of large-scale farming methods on soil exploitation in Czechoslovakia.
3. Effects of intensification of agriculture on nature and landscape in the Netherlands.
4. Saline seeps in northern Great Plains, the United States, and Canada.
5. Soil erosion and degradation in southern Piedmont of the United States.
6. U.S. soil depletion study of the southern Iowa River basin.
7. Reclamation of areas affected by open-cast mining in Czechoslovakia.
8. Transformation of small villages into rural cities in Czechoslovakia.

3.1.2 *The Earth as Transformed by Human Action* (Turner et al., 1989)

This book is a compilation of papers presented at the "Earth as Transformed by Human Action" symposium held at Clark University in 1987 as part of

Clark's centennial-year celebration. The book and symposium were the results of an ambitious effort whose goals were to document changes in the biosphere over the past 300 years, to contrast the global patterns of change with those experienced at the regional level, and to explore the major human forces that have driven changes in the biosphere.

A thorough review of this book was not undertaken in this document due to time constraints. A summary of the book, quoted from its preface, is given below.

> The text is composed of an introduction and four principal sections. The introductory chapter of the volume establishes the intellectual ancestry of the subject of *The Earth as Transformed by Human Action* and briefly traces some of the basic views of the human-nature relationships of the last 300 years. It then summarizes the major findings of the volume as a whole, assessing the major trends in the transformation variables and the major patterns found in the regional case studies.
>
> Section I, *Changes in Population and Society*, examines five major human forces of change over the past 300 years: population, technology, institutions/organization/culture, location of production and consumption, and urbanization. The stage for these five studies is set by the lead chapter of the section, which examines long-term, regional population changes, and the section is set in intellectual context by a concluding chapter on the history of beliefs regarding transformation, which themselves may also be seen as real or potential human forces of environmental change.
>
> Section II, *Transformations of the Global Environment*, consists of 18 papers that address the principal objective of the volume, a stocktaking of the major transformations of the biosphere wrought by human action over the past 300 years. Again, the first chapter of the section establishes the context by assessing long-term changes in the biosphere of natural origin. The other papers attempt to track the changes in the components of the biosphere, either a single variable of a set of variable. These are arranged in subsections: land, water, oceans and atmosphere, biota, and chemicals and radiation.
>
> Section III, *Regional Studies of Transformations*, is comprised of 12 case studies that document the multiple-variable interactions of environmental change over a 300-year period for specific areas, serving as spatial and conceptual comparisons for the global papers.
>
> Section IV, *Understanding Transformations*, briefly examines a range of perspectives and theories that purport to explain human actions in regard to the biosphere. Three papers address such themes as they emanate from the realms of meaning, social relations, and ecology.

3.1.3 Studies of Deforestation/Reforestation

3.1.3.1 "Deforestation" (Arnold, 1987). This article was prepared for the Dahlem Workshop on *Resources and World Development*. It provides a short overview on deforestation.

Data. Deforestation estimates are based on data from the 1980 FAO/UNEP study of tropical forest resources (Lanly, 1982).

Process. The principal causes of deforestation in the tropics are rapid population growth coinciding with poverty, unequal distribution of land, and low agricultural productivity. In closed tree formations, shifting cultivation is the principal cause of deforestation for all regions. Grazing is the second most important cause. Timber harvesting is an important cause in Asia and Africa. This is because logging roads open the area for agriculture. In open tree formations, shifting cultivation and grazing are the major causes of deforestation. Fuel wood harvesting is another important cause.

This article also discusses the consequences of deforestation (ecological, social, and economic) and policies for reducing deforestation.

3.1.3.2 *Global Deforestation and the Nineteenth Century World Economy* (Tucker and Richards, 1983) and *World Deforestation in the Twentieth Century* (Richards and Tucker, 1988). These books are based on case studies presented at two separate symposia. The goal of both meetings was to draw generalizations and themes from a diverse set of studies from around the globe. The essays in these books document, describe, and analyze aspects of the world trend toward deforestation.

Data. Both books are mainly a compilation of case studies. Several of the authors in the book on the twentieth century address the problem of collecting adequately detailed, accurate, comparable data across time and space.

Process. The dominant cause of deforestation in the nineteenth century was "the steeply rising demand for production of agricultural commodities exerted by the core or metropolitan societies of Europe, North America, and Japan" (p. xi). A dominant theme in these essays was the increasing unification of the global economy under the leadership of British capital, technology, and imperial institutions.

> The global economy continues to be a significant factor in the twentieth century. Thus, in addition to rural population growth as a factor in deforestation (through demands on timber and land resources for agriculture), the impact of industrial economies remains a critical contributing factor.
>
> We see more clearly the impact of outside capital: industrial economies tapping the developing economies' timber resources to meet their consumption demand and private investors (in some cases in alliance with local commercial interests) cashing in on the high short-term profitability of timber exports from capital-starved countries. The consequences of this imbalance of power between industrialized and developing nations are the main concern of this volume. (p. 4)

In the volume on the twentieth century, the themes explored in case studies include the ability of some industrialized countries to reverse the trend of deforestation; the relationship between timber as a commodity and deforestation; interactions between Western capital and regional markets controlled by local entrepreneurs responding to regional opportunities for profit (control of timber by international interests); the role of development policies; the role of the timber lobby; the disruption of traditional production and social systems; and the effect of modern forest management on land and indigenous people.

The case studies demonstrate that until very recently an awareness of environmental costs has occurred only in industrialized, high-literacy, high-income countries.

3.1.3.3 *Conversion of Tropical Moist Forests* (Myers, 1980) and *The Primary Source* (Myers, 1984). The report *Conversion of Tropical Moist Forests* was commissioned by the National Research Council's Committee on Research Priorities in Tropical Biology. The goal was to document the forms and degree of tropical forest destruction. *The Primary Source* is an update of that survey.

Data. This document makes clear the shortcomings in data on the amount of deforestation. It provides a review of forest resources and the rates of forest depletion (deforestation and degradation) for tropical countries. It classifies areas as undergoing rapid rates of conversion, moderate rates, or little change.

Process. This examines the role of the following factors in deforestation: forest farmers, the timber trade, cattle raising, and firewood cutting. Population pressure, particularly from forest farmers, is identified as the greatest factor in deforestation.

3.1.3.4 *Quantifying Changes in Forest Cover in the Humid Tropics: Overcoming Current Limitations* (Grainger, 1984). This work develops a model of deforestation based on population, food consumption per capita, and average yield per hectare. The model is used to forecast deforestation for 43 tropical countries. A major assumption it that once forest area has fallen to a critical level in a given country, the government will take action to prevent further deforestation. This work is cited in *World Resources 1988-1989* (World Resources Institute and International Institute for Environment and Development, 1988). It has not been reviewed by this author.

3.1.3.5 "Deforestation Perspectives for the Tropics: A Provisional Theory with Pilot Applications" (M. Palo, 1987). This is a chapter in *The Global Forest Sector*, the IIASA study that is reviewed in section 3.4.5.

Data. Although actual case studies are not presented in this chapter, the model is primarily based on observations from field work in four countries.

Process. This paper develops a theory of the causes of deforestation in the tropics. This is a systematic, interdisciplinary, global theory consisting of 20 hypotheses. The theory takes into account natural factors, accessibility, population pressures, public ownership (public goods), government policies, colonialism, and positive feedback loops that increase the rate of deforestation.

3.1.3.6 Other Case Studies. There are many case studies on deforestation. In addition to those mentioned above: "Borneo and Peninsular Malaysia" in *The Earth as Transformed by Human Action*, Turner et al., eds. (Brookfield et al., 1990).

3.1.4 Studies of Wetland Conversion

3.1.4.1 "Forested Wetland Depletion in the United States: An Analysis of Unintended Consequences of Federal Policy and Programs" (Stavins and Jaffe, 1988), and "Alternative Renewable Resource Strategies: A Simulation of Optimal Use" (Stavins, 1989). These two papers develop an economically driven model of the conversion between forested wetlands and farmland in the Lower Mississippi Alluvial Plain. They are reviewed in section 3.3.1.

3.1.4.2 Case Studies. There are many case studies of wetland conversion, including:

- "Sweden" in *The Earth as Transformed by Human Action* (Hagerstrand and Lohm, 1990).
- "The Impact of Wetland Reclamation" in *Land Transformation in Agriculture*, case studies of Indonesia and China (Ruddle, 1987).
- *The Changing Fenlands* (Darby, 1983).
- "Drainage and Economic Development of Poles'ye, USSR" (French, 1959).
- "The Reclamation of Swamp in Pre-Revolutionary Russia" (French, 1964).
- "Draining the Swamps" in *The Making of the South Australian Landscape* (Williams, 1974).

3.1.5 Studies of Burning

3.1.5.1 "The Role of Fire" (Robinson, 1987). This dissertation provides an in-depth overview of fire as a force in transforming the landscape and as a contributor to the chemical and radiative behavior of the atmosphere. A large portion of this work is devoted to evaluating the prospects for using remote sensing and related information systems for assessing the role of fire on earth.

Data. This study provides a critical review of current estimates of emissions from burning. It suggests where and why these estimates are most in error (see also Robinson, 1986). It provides a detailed review of aerosol and trace gas emissions from burning, with a compilation of emission coefficients and global estimates. It presents data on calculated global surface type and albedo changes. Five case studies of burning are presented: Lago Calado, Amazonas, Brazil; Transamazon. Km 50. Para, Brazil; Chiapas, Mexico; Minas Gerais, Brazil; and Hengchun, Taiwan.

This study notes that while agricultural and cooking fires are responsible for a large fraction of total biomass burned, data on these activities are quite poor. Satellite remote sensing is unlikely to improve these estimates. Robinson suggests that social indices may be the most useful approach for inferring the magnitude of burning (Robinson, 1987, pp. 285-293).

Process. Agricultural and cooking fire regimes are closely related to many social factors, including population pressures, surplus labor, poverty, and agricultural practices. This work discusses the relationship between anthropogenic fire regimes and population density (pp. 285-293).

3.1.5.2 *Fire in America, A Cultural History of Wildland and Rural Fire* (Pyne, 1982). This book provides a chronological history of fire in America, as well as detailed regional histories. It discusses fire as a cultural phenomenon, as an environmental modifier, and in relationship to social organization. "The relationship between mankind and fire is reciprocal: fire has made possible most technological and agricultural developments and has provoked fundamental intellectual discourse; yet fire itself takes on many particular characteristics because of the cultural environment in which it occurs, just as it does in response to the natural environment of fuels, topography and weather. . . . And it is the culture of fire—as distinct from its physics, chemistry, biology, and meteorology—that forms the subject of this study." (p. 5)

3.1.5.3 "Estimation of Gross and Net Fluxes of Carbon between the Biosphere and the Atmosphere from Biomass Burning" (Seiler and Crutzen, 1980).

This paper estimates global CO_2 releases from biomass burning. There is a large range in the estimates of carbon flux from burning. Other studies that estimate current emissions from all land use changes include Houghton et al. (1987), Detwiler and Hall (1988), and Bolin et al. (1986). Estimates of cumulative rates of CO_2 release from deforestation have been made by Woodwell et al. (1983) and Bolin et al. (1986).

Data. The following activities that lead to burning were included in the calculations: tropical shifting agriculture, deforestation due to population increase and development programs, industrialization and colonization, natural or agricultural fires in savanna areas, wildfires in temperate forests, prescribed fires, wildfires in boreal forests, burning of industrial wood and fuel wood, and burning of agricultural wastes. Estimates on the level of these activities are made from a variety of data sources. The authors recognize the large uncertainty in their estimates due to limitations in the data on which the estimates were made.

Process. Estimates of biomass burned were made using the following model:

$$M = A * B * C * D$$

where

A = total land area burned annually,

B = average organic matter per unit area,

C = fraction of the average above-ground biomass relative to total average biomass, and

D = burning efficiency of above-ground biomass.

The parameters A through D were generally estimated by a critical review of values cited in the literature.

Synthesis. This study used measures of the level of human activities in combination with technical information about emissions from burning to determine the carbon flux.

3.2 Technological (and Institutional) Change in Agriculture

3.2.1 *Agricultural Development* (Hayami and Ruttan, 1985)

This book develops the theory of induced innovation as an explanation of alternative paths and rates of agricultural development.

Data. This study includes cross-section data on levels of production, productivity, and inputs in agriculture for the years 1960 and 1980 (44 countries); time series data on the United States and Japan for the years 1880 to 1980; and data at the village level for the Philippines and Indonesia.

Process. A brief summary of the theory of induced innovation in agriculture is given as follows: Technological change is mostly endogenous, induced by changes in relative resource endowments (factor supply) and the growth of demand (product demand). In this scheme, improvements in mechanization provide substitutes for labor; biological and chemical innovations provide substitutes for land. Institutional innovation is induced both by changes in relative resource endowments and by technical change. The theory and analysis concentrate on resource endowments, technological change, and institutional change.

The econometric analysis presented in this book supports the theory of induced innovation. For an abbreviated discussion of this analysis, see Ruttan (1985). Further development of the ideas of induced innovation can be found in Binswanger et al. (1978) and Ahmed and Ruttan (1988).

3.3 Dynamic Land Use Modeling

Dynamic land use modeling describes how land is allocated over time between competing uses. An article by Parks and Alig (1988) reviews models of land use conversion at the regional level. There are no dynamic models of land use on the global scale.

This review article presents a taxonomy of land use modeling approaches:

(1) inventory and descriptive studies that classify the physical amount and characteristics of land or its subclasses; (2) normative, optimizing models that explain how land should be used in relation to various objectives; (3) positive studies that explain the use of land as it relates to economic, social, policy, climatic, and other variables.

Positive and normative models are generally based on neoclassical economic theory.

3.3.1 "Forested Wetland Depletion in the United States: An Analysis of Unintended Consequences of Federal Policy and Programs" (Stavins and Jaffe, 1988), and "Alternative Renewable Resource Strategies: A Simulation of Optimal Use" (Stavins, 1989)

These two discussion papers examine the conversion between forested wetlands and farmland in the lower Mississippi alluvial plain. The major contribution of this work is the development of a model based on the heterogeneity of the land base with parameters estimated from land use data.

Data. This work used data for the years 1935 to 1984. Data on forested land was based on U.S. Forest Service aerial photographs. The study also employed data on agricultural revenue, agricultural costs of production, forestry revenue, flood and drainage conditions, flood and drainage protection, weather conditions, and costs of conversion.

Process. This study develops a model of land use based on individual firms making rational economic decisions. Land use decisions are modeled as a function of the relative expected economic returns from alternative land uses. The study examines the effect of federal flood control strategies and price changes. The study also develops a model for the socially optimal time-path of resource use that is analogous to the model for the individual firm, but includes the cost of externalities in the conversion of wetlands to cropland.

Synthesis. A model of the heterogeneity of the land base was estimated from land use data. This model was integrated with an econometrically estimatable model of the effect of economic and policy variables.

3.4 Agricultural and Forestry Production Modeling

This section reviews several global models of agriculture production. Some of the agricultural models assess the productive capacity of the world's land base, while others look at the supply and demand for agricultural products. These models only project scenarios to the year 2000 or 2010. There are currently no agricultural models that provide an agro-ecological assessment, incorporate economic processes, and look at a time horizon 80 to 100 years into the future. In addition, the existing models are not dynamic with regard to changes in the land base. The situation for forestry is not much different. The global forestry model reviewed in this section was developed to understand supply and demand of wood products, and does not illuminate land use changes or other environmental effects of forest use.

The above discussion points to the fact that by the definition used for this study, these models do not provide a "synthetic" component. It is for this reason that the reviews of this section do not include the category "synthesis." This does not imply that these complex models did not require the synthesis of a significant range of concepts, analytical techniques, and data.

A detailed review of agriculture models, with a focus on the ability of these models to illuminate relationships between development and environment, is found in *Scenarios of Socioeconomic Development for Studies of Global Environmental Change: A Critical Review* (Toth et al., 1989). Another comprehensive review of these models, with a focus on the relationship between population growth and food, is found in Srinivasan (1988). A review focused on the less developed countries' food balance is found in Fox and Ruttan (1983).

APPENDIX B

3.4.1 Agro-ecological Zones (AEZ) Project (Food and Agriculture Organization (FAO) of the United Nations, 1978-1981; Shah et al., 1985)

The goal of this study was to assess the rainfed production potential and the population supporting capacity of the world land resources.

Data. The study develops a data base of climate and soil characteristics for the developing world, with spatial disaggregation of 50,000 land units and 14 major climates. The model results provide data on the land base by region and possible losses to the land base due to erosion if conservation practices are not used. The study also calculates per capita calorie and protein requirements for present and future populations. It includes a detailed case study of Kenya.

Process. The study incorporates the developing world only. The model assesses climate and soil characteristics. Based on climate and soil suitability, the land suitability by crop for three levels of technology is determined. From this, the potential crop yield is calculated. The level of irrigated production and demand for food per capita are exogenous variables. The land loss due to erosion when no conservation practices are employed is assessed. Except for soil loss and productivity loss, the model is static.

3.4.2 *Model of International Relations in Agriculture (MOIRA)* (Linnemann et al., 1979)

This study evaluates the production potential for food and policy options for ameliorating world hunger. It contains both a food production potential model and an economic model.

Data. For soil assessment, the world is divided into 222 land units. For economic assessment, the world is divided into 106 geographical units. Country-level data for the year 1965 includes the ratio of non-food to food agricultural production, ratio of non-agriculture to agriculture per capita incomes, sectoral income distribution, fish catch and distribution, and technological parameters in agriculture.

Process. In the food production model, this study determines the availability of agricultural land and applies a theoretical maximum rate of photosynthesis. The food production potential model is static. The economic model is based on an economic equilibrium model with international trade in food. Consumption and production are dependent on the domestic food price, which is subject to government intervention. Regression analysis, based on 1965 country-level data, is used to determine parameters in the model. These structural parameters do not change. The model assumes values at the country level for three exogenous variables: population growth, non-agricultural gross domestic production growth, and regional fertilizer prices. Projections are made for 1-year periods to the year 2010.

3.4.3 Exploring National Food Policies in an International Setting (Parikh, 1981)

The goals of the Food and Agricultural Program at IIASA were to evaluate the world food situation, to identify underlying factors, and to suggest policy alternatives at the national, regional, and global levels. The basic linked system was one of the products of this effort.

Data. Model parameters are calibrated based on the FAO's supply utilization accounts for the years 1970 to 1976.

Process. This is a dynamic general equilibrium model. It includes 18 country models, 2 country group models, and 14 regional group models that are linked together in trade, aid, and capital flows. Projections are made to the year 2000. The model is structured to evaluate the effect of policy alternatives on output. The policies examined include domestic price policies, quantity rationing, trade restrictions, strategic reserves, normative consumption and import, plan target realization, self-sufficiency, and free market on output. A major shortcoming of this model for long-term studies is that available land is treated as a time trend.

This model has been modified to extend the time horizon to the year 2050 for use in the U.S. EPA Stabilization Study.

3.4.4 Other Agriculture Models

Several economic models simulate supply, demand, distribution, and hunger. These include the following:

- *Agriculture—Toward 2000* (Food and Agriculture Organization (FAO) of the United Nations, 1981). This is a normative model that forecasts to the year 2000.
- *Resources and Environmental Effects of U.S. Agriculture* (Crosson and Sterling, 1982) and *Global 2000 Report to the President* (Council on Environmental Quality and the Department of State, 1980). These two studies provide good information on the effects of agriculture on the environment. They are not useful on a global scale for production and consumption questions. The Crosson and Sterling (1982) study looks at U.S. production only.

APPENDIX B

3.4.5 *The Global Forest Sector* (Kallio et al., 1987)

This is the final report of the Forest Sector Project at the International Institute for Applied Systems Analysis (IIASA). The goals of this project were to study long-term developments in the production, consumption, and world trade of forest products and to develop a policy analysis tool. The study focuses on the use of wood, and not other benefits of forests. It presents a detailed discussion of the global forest sector model developed at IIASA. This volume also provides a thorough critical review of modeling approaches for supply, demand, and trade in the forest sector.

Data. Many different contributors to this volume noted the lack of consistent comparable data on forest resources as a major obstacle to improved modeling in this sector. Improvement in data is more important than improvement in estimation techniques. Better data is needed on stumpage prices, harvest qualities, forest characteristics, and ownership variables.

This modeling effort relied heavily on existing data that is referenced throughout the work.

Process. The global forest sector model is a partial equilibrium model using a nonlinear programming framework. The model has four modules: timber supply, forest products industry, product demand, and international trade. The world is divided into 18 regions. There are 16 forest products. The planning horizon is 50 years. The model is designed for evaluating future scenarios with differing assumptions about socioeconomic and environmental factors.

The forest resource is modeled by a simple growth function, with parameters estimated from historical data where feasible. The changes in the land area used for forests (afforestation and deforestation) were estimated exogenously.

Chapter 3 develops a theory of the causes of deforestation in the tropics. This is reviewed in section 3.1.3.5.

4 DATA AND MONITORING

There are significant data requirements for developing a better understanding of human interactions with global change. A thorough review of existing data sources is beyond the scope of this paper.[1] Instead, this section strives to be illustrative by highlighting several examples of data bases that were explicitly developed to improve our understanding of global environmental change.[2] Although limited in scope, the goal of this section is to act as a catalyst for thinking about what new data are most needed to develop a better understanding of the processes of change in the two areas of concern to this committee: land use transformations and industrial transformations of materials and energy.

The data bases listed below are of two kinds: (1) data on levels of human activity (e.g., deforestation) and (2) quantitative data on emissions that are calculated based on the level of human activity and information on emission factors (e.g., CO_2 emissions from deforestation). A general knowledge of a broad class of economic and social data bases is assumed and thus not reviewed here.

- Emissions of CH_4, NO_x, SO_x, HCl, and sea salt on a regional basis, and CH_4, CO, NO_x, N_2O, and CFCs on a global basis for the years 1800 to 1980, in 30-year intervals, excluding 1830 (Darmstadter et al., 1987). The study contributes new data in historical estimates of land in wet rice cultivation, and for emissions from combustion, the flaring of natural gas, smelters, cokers, and other industrial processes. For a more in-depth review, see section 2.1.2.
- Current (mean value for 1980 to 1986) and cumulative (for years 1860 to 1986) releases of CO_2 from fossil fuel combustion and biota for most countries of the world (Subak, 1989). Estimates for biota are "fairly crude" because data on deforestation and biomass burning are not yet well documented.
- Annual CO_2 emissions from fossil fuels, by country, for the years 1949 to 1986 (Marland et al., 1988). Based on U.N. energy statistics.
- Annual global emissions of CO_2 from fossil fuel combustion for the years 1860 to 1982 (Rotty and Masters, 1985).
- CO_2 releases from land clearing for agricultural purposes, for the years 1860 to 1986 (Richards et al., 1983).
- Energy consumption by end use sector for all countries (Mintzer, 1988, for industrialized countries; Sathaye et al., 1988, for developing countries).
- Forest resources, and amount and rates of deforestation for the 1980s, by country (IIED and WRI, 1987). Data are based on the U.N. Food and Agriculture Organization, the U.N. Economic Commission for Europe, and country data sources.
- Forest resources and the rates of deforestation and forest degradation for tropical countries (Myers, 1980, 1984). For a review of this work, see section 3.1.3.3.
- Data on production of halocarbons from 1960 to 1985 (U.S. EPA, 1987; Hammit et al., 1986).
- Global anthropogenic emissions of trace metals to the atmosphere, water, and soil (Nriagu and Pacyna, 1988). Data on emission factors for key anthropogenic processes.
- Natural emissions of trace metals to the atmosphere and comparison of natural and anthropogenic emissions to atmosphere (Nriagu, 1989).

5 GLOBAL POPULATION MODELS

In the models reviewed in the body of this report, population is always specified exogenously. Population estimates are generally derived from one of a few models, which will be described below. These models tend to have similar estimates to the year 2025, with some divergence when projecting further into the future. For a review of population models with a focus on the ability of these models to illuminate relationships between development and environment, see Toth et al. (1989). For a critical review of global population modeling, see Keyfitz (1981, 1982) and Lee (1989).

The most widely used models for forecasting and scenario development have much in common. The key parameters in population models are initial population size and age-sex structure, fertility rates, mortality rates, and net migration rates. Estimates of fertility rates are the greatest source of uncertainty in these models. Determination of the values for key parameters in population models is based on one of two approaches: (1) trend extrapolation, modified by expert judgment, or (2) assuming a date in the future when replacement-level fertility will be reached, and using linear interpolation to determine intervening rates. Both of these methods are based on expert judgment. There is no clear theoretical explanation on which population models are built.

In concluding his review, Lee (1989) emphasized the lack of consistent theory behind long-term global population forecasts.

> Current longrun population forecasts ignore economic, natural resource and environmental constraints. Yet they assume that populations are even now converging to stationarity at a global level about twice the current population. If the assumption derives from a Malthusian orientation, it must be based on unexpressed and, in this context, unexamined views about future growth prospects and reproductive response to economic or environmental change. . . .
>
> If, instead, population convergence to stationarity has been inferred from some version of transition theory, such as modern socio-economic fertility models, then again the forecasts rest on unexamined assumptions. They must assume that growth and development will proceed along global trend patterns without encountering serious Malthusian constraints. . . . The assumption that the end point of the transition is at replacement level fertility is supported neither by history nor by the logic of relevant social theory.

A review of global population models (Toth et al., 1989) recommended three models as most suitable for use in long-term, large-scale development-environment studies:

1. *World Population Prospects Estimates and Projections as Assessed in 1982* (United Nations, 1985).

2. "Global Population (1975-2075)" and "Labor Force (1975-2050)" (Keyfitz et al., 1983).

3. *World Development Report 1984, World Population Projections 1984* (World Bank, 1984).

NOTES

1. For current information on sources of data describing greenhouse gas emission levels, and of human activities that cause greenhouse gas emissions, see Lashof and Tirpak (1989). For a discussion of the strengths and weaknesses of current observational programs in the area of human interactions with global environmental change, see Committee on Earth Sciences (CES, 1989).

2. The examples given are based on the author's knowledge and do not represent a thorough review of all data. Lack of a listing does not necessarily indicate there are no appropriate data bases. Likewise, inclusion does not indicate reliability of the data.

REFERENCES AND SELECTED READING

Ahmed, I., and V.W. Ruttan (eds.). 1988. Generation and Diffusion of Agricultural Innovations: The Role of Institutional Factors. Gower Publishing Company Limited, Aldershot, England.

Anderberg, S. 1989. A conventional wisdom scenario for global population, energy, and agriculture 1975-2075, and surprise-rich scenarios for global population, energy and agriculture 1975-2075. In F.L. Toth et al. (eds.), Scenarios of Socioeconomic Development for Studies of Global Environmental Change: A Critical Review. RR-89-4. International Institute for Applied Systems Analysis, Laxenburg, Austria.

Arnold, J.E.M. 1987. Deforestation. In D.J. McLaren and B.J. Skinner (eds.), Resources and World Development. John Wiley and Sons, Chichester, England.

Ausubel, J.H. 1989. Regularities in technological development: An environmental view. In J.H. Ausubel and H.E. Sladovich (eds.), Technology and Environment. National Academy Press, Washington, D.C.

Ausubel, J.H., and R. Herman (eds.). 1988. Cities and Their Vital Systems: Infrastructure Past, Present, and Future. National Academy Press, Washington, D.C.

Ausubel, J.H., A. Grubler, and N. Nakecenovic. 1988. Carbon dioxide emissions in a methane economy. Climatic Change 12(3):241-265.

Ayres, R.U. 1989a. Industrial Metabolism in Technology and Environment. J.H. Ausubel and H.E. Sladovich (eds.), Technology and Environment. National Academy Press, Washington, D.C.

Ayres, R.U. 1989b. Technological Transformations and Long Waves. Research Report 89-1. International Institute for Applied Systems Analysis, Laxenburg, Austria.

Ayres, R.U., and S.R. Rod. 1986. Reconstructing an environmental history: patterns of pollution in the Hudson-Raritan Basin. Environment 28(4):14-20, 39-43.

Ayres, R.U., L.W. Ayres, J.A. Tarr, and R.C. Widgery. 1988. An Historical Reconstruction of Major Pollutant Levels in the Hudson-Raritan Basin: 1880-1980. NOAA Technical Memorandum NOS OMA 42. National Oceanic and Atmospheric Administration, United States Department of Commerce, Washington, D.C.

Ayres, R.U., V. Norberg-Bohm, J. Prince, W.M. Stigliani, and J. Yanowitz. 1989. Industrial Metabolism, the Environment, and Application of Materials-Balance Principles for Selected Chemicals. RR-89-11. International Institute for Applied Systems Analysis, Laxenburg, Austria.

Binswanger, H.P., et al. 1978. Induced Innovation. John Hopkins University Press, Baltimore.

Bolin, B., B.R. Doos, J. Jager, and R.A. Warrick (eds.). 1986. The Greenhouse Effect, Climatic Change, and Ecosystems. Scope 29. John Wiley and Sons, Chichester, England.

Brookfield, H.C., F. Lian, L. Kwai-Sim, and L. Potter. 1990. Borneo and peninsular Malaysia. In B.L. Turner et al. (eds.), The Earth as Transformed by Human Action. Cambridge University Press, New York.

Brower, F.M., and M.J. Chadwick. 1988. Future Land Use Patterns in Europe. IIASA WP-88-040. International Institute for Applied Systems Analysis, Laxenburg, Austria.

Burke, L.M., and D.A. Lashof. 1989. Greenhouse Gas Emissions Related to Agriculture and Land-Use Practices. Prepared for the Annual Meeting Proceedings of the Agronomy Society of America, Nov. 27 to Dec. 2, 1988, Anaheim, Calif. (supported by U.S. Environmental Protection Agency).

Chandler, W.U. 1988. Assessing the carbon emission control strategies: The case of China. Climatic Change 13(3):241-265.

Committee on Earth Sciences (CES). 1989. Our Changing Planet: The FY 1990 Research Plan. Federal Coordinating Council on Science, Engineering, and Technology. Office of Science and Technology Policy, Washington, D.C.

Council on Environmental Quality and the Department of State. 1980. Global 2000 Report to the President: Entering the Twenty-first Century. (three volumes.) U.S. Government Printing Office, Washington, D.C.

Crosson, P.R., and B. Sterling. 1982. Resources and Environmental Effects of U.S. Agriculture. Research paper. Resources for the Future, Washington, D.C.

Crutzen, P.J. 1987. Role of the tropics in atmospheric chemistry. In R. Dickinson (ed.), Geophysiology of Amazonia. John Wiley and Sons, New York.

Darby, H.C. 1983. The Changing Fenlands. Cambridge University Press, New York.

Darmstadter, J., L.W. Ayres, R.U. Ayres, W.C. Clark, P. Crosson, P.J. Crutzen, T.E. Graedel, R. McGill, J.F. Richards, and J.A. Tarr. 1987. Impacts of World Development on Selected Characteristics of the Atmosphere: An Integrative Approach. Vols. 1 and 2. ORNL/Sub/86-22033/1/V1. Oak Ridge National Laboratory, Oak Ridge, Tenn.

Detwiler, R.P., and C.A. Hall. 1988. Tropical forests and the global carbon cycle. Science 239:42-47.

Duchin, F. 1988a. Analyzing structural change in the economy. In M. Ciaschini (ed.), Input-Output Analysis: Current Developments. Chapman Hall, London.

Duchin, F. 1988b. Analyzing Technological Change: An Engineering Data Base for Input-Output Models of the Economy. Engineering with Computers 4:99-105.

Duchin, F. 1989a. An input-output approach to analyzing the future economic implications of technological change. In R. Miller, K. Polenske, and A. Rose (eds.), Frontiers of Input-Output Analysis. Oxford University Press, New York.

Duchin, F. 1989b. Framework for the Evaluation of Scenarios for the Conversion of Biological Materials and Wastes to Useful Products: An Input-Output Approach. Presented at the joint session of the American Economics Association/American Association for the Advancement of Science, Prospects and Strategies for the American Economy, ASSA Meetings, December 29, 1988, New York.

Duchin, F. 1989c. Project Proposal: Strategies for Environmentally Sound Development: An Input-Output Analysis. Institute for Economic Analysis, New York University, New York.

Edmonds, J. 1988. Editorial response to Bill Keepin. Climatic Change 13(3):237-240.

Edmonds, J. 1989. A Second Generation Greenhouse Gas Emissions Model. Outline of Design and Approach. Pacific Northwest Laboratory. Washington, D.C.

Edmonds, J., and J. Reilly. 1983. Global Energy and CO_2 to the Year 2050. The Energy Journal 4(3).

Edmonds, J., and J. Reilly. 1985a. Future global energy and carbon dioxide emissions. In J.R. Trabalka (ed.), Atmospheric Carbon Dioxide and the Global Carbon Cycle. DOE/ER-0239. U.S. Department of Energy, Washington, D.C.

Edmonds, J., and J. Reilly. 1985b. Global Energy: Assessing the Future. Oxford University Press, New York.

Fisher, G., K. Frohberg, M.A. Keyzer, and K.S. Parikh. 1988. Linked National Models: A Tool For International Food Policy Analysis. Kluwer, Dordrecht, The Netherlands.

Fisher, J.C., and R.H. Pry. 1970. A simple substitution model of technological change. Technol. Forecast. Soc. Change 3:75-88.

Food and Agriculture Organization (FAO) of the United Nations. 1978-1981. Reports of the Agro-ecological Zones Project. World Soil Resources Report 48. Vols. 1 to 4. FAO, Rome.

Food and Agriculture Organization (FAO) of the United Nations. 1981. Agriculture—Toward 2000. Economic and Social Development Series, 23. FAO, Rome.

Fox, G., and V.W. Ruttan. 1983. A guide to LDC food balance projections. European Review of Agricultural Economics 10(4):325-356.

French, R.A. 1959. Drainage and Economic Development of Poles'ye, USSR. Economic Geography 25:172-180.

French, R.A. 1964. The reclamation of swamp in pre-revolutionary Russia. Transactions and Papers of British Geographers 34:175-188.

Frohberg, K.K., and P.R. Van de Kamp. 1988. Results of Eight Agricultural Policy Scenarios for Reducing Agricultural Sources of Trace Gas Emissions. Office of Policy Analysis, U.S. Environmental Protection Agency, Washington, D.C.

Frosch, R., J.H. Ausubel, and R. Herman. 1989. Technology and environment: An overview. In J.H. Ausubel and H.E. Sladovich (eds.), Technology and Environment. National Academy Press, Washington, D.C.

Goldemberg, J., T.B. Johansson, A.K.N. Reddy, and R.H. Williams. 1985. An end-use oriented global energy strategy. In Annual Review of Energy 1985. Annual Reviews Press, Palo Alto, Calif.

Goldemberg, J., T.B. Johansson, A.K.N. Reddy, and R.H. Williams. 1987. Energy for a Sustainable World. World Resources Institute, Washington, D.C.

Goldemberg, J., T.B. Johansson, A.K.N. Reddy, and R.H. Williams. 1988. Energy for a Sustainable World. Wiley Eastern Limited, New Delhi.

Gordon, R.B., T.C. Koopmans, W.D. Nordhaus, and B.J. Skinner. 1987. Toward a New Iron Age. Harvard University Press, Cambridge, Mass.

Grainger, A. 1984. Quantifying changes in forest cover in the humid tropics: Overcoming current limitations. Journal of World Forest Resource Management 1(1):3-63.

Grainger, A. 1987. A land use simulation model for the humid tropics. Proceedings of International Conference on Land and Resource Evaluation for National Planning in the Tropics, January 25-31, 1987, Chetumal, Mexico. Forest Service, U.S. Department of Agriculture, Washington, D.C.

Haefele, W., and P. Basile. 1979. Modelling of long range energy strategies with a global perspective. Pp. 493-529 in K.B. Haley (ed.), Operation Research '78. North Holland, Amsterdam.

Haefele, W., J. Anderer, A. McDonald, and N. Nakicenovic. 1981. Energy in a Finite World. Vol. 1: Paths to a Sustainable Future. Vol. 2: A Global System Analysis. Report by the Energy Systems Program Group. Ballinger, Cambridge, Mass.

Hagerstrand, T., and U. Lohm. 1990. Sweden. In B.L. Turner et al. (eds.), The Earth as Transformed by Human Action. Cambridge University Press, New York.

Hammit, J.K., K.A. Wolf, F. Camm, W.E. Mooz, T.H. Quin, and A. Bamezai. 1986. Product Uses and Market Trends for Potential Ozone-Depleting Substances. U.S. Environmental Protection Agency and RAND, Santa Monica, Calif.

Hayami, Y., and Ruttan, V.W. 1985. Agricultural Development. The Johns Hopkins University Press, Baltimore.

Heany, J.P., W.C. Huber, and S.J. Nix. 1976. Storm water management model: Level 1, Preliminary screening procedures. EPA-600/2-76-275. U.S. Environmental Protection Agency, Washington, D.C.

Herman, R., S.A. Ardekani, and J.H. Ausubel. 1989. Dematerialization. In J.H. Ausubel and H.E. Sladovich (eds.), Technology and Environment. National Academy Press, Washington, D.C.

Houghton, R.A., R.D. Boone, J.E. Fruci, J.E. Hobbie, J.M. Melillo, C.A. Palm, B.J. Peterson, G.R. Shaver, G.M. Woodwell, B. Moore, D.L. Skole, and N. Myers. 1987. The flux of carbon from terrestrial ecosystems to the atmosphere in 1980 due to changes in land use: Geographic distribution of the global flux. Tellus 39B:122-139.

International Institute for Environment and Development (IIED) and World Resources Institute (WRI). 1987. World Resources 1987. Basic Books, New York.

Kallio, M., D.P. Dykstra, and C.S. Binkley (eds.). 1987. The Global Forest Sector. John Wiley and Sons, Chichester, England.

Keepin, B. 1988. Caveats in global energy/CO_2 modeling. Climatic Change 13(3):233-235.

Keyfitz, N. 1981. The limits of population forecasting. Population and Development Review 7(4):579-593.

Keyfitz, N. 1982. Can knowledge improve forecasts? Population and Development Review 8(4):729-751.

Keyfitz, N., E. Allen, J. Edmonds, R. Doughes, and B. Wiget. 1983. Global population (1975-2075) and labor force (1975-2050). Institute for Energy Analysis, Oak Ridge Associated Universities. ORAU/IEA-83-6(M). Oak Ridge, Tenn. 67 pp.

Lanly, J.P. 1982. Tropical Forest Resources. United Nations Food and Agriculture Organization, Rome.

Lashof, D.A., and D.A. Tirpak. 1989. Policy Options for Stabilizing Global Climate. Draft Report to Congress. Office of Policy, Planning, and Evaluation, U.S. Environmental Protection Agency, Washington, D.C.

Lee, R. 1989. The Second Tragedy of the Commons. Graduate Group in Demography, University of California, Berkeley.

Lee, R. undated. Longrun Global Population Forecasts: A Critical Appraisal. Demography and Economics, University of California, Berkeley.

Lee, T.H., and N. Nakicenovic. 1988. Technology Life-Cycles and Business Decisions. International Journal of Technology Management 3(4):411-426.

Linnemann, H., J. De Hoogh, M.A. Kayzer, and H.D.J. Van Heemst. 1979. Model of International Relations in Agriculture (MOIRA). North-Holland, Amsterdam.

Manning, E.W. 1988. The Analysis of Land Use Determinants in Support of Sustainable Development. International Institute for Applied Systems Analysis, Laxenburg, Austria.

Marchetti, C. 1981. Society as a Learning System: Discovery, Invention, and Innovation Cycles Revisited. RR-81-29. International Institute for Applied Systems Analysis, Laxenburg, Austria. November.

Marchetti, C. 1983. The Automobile in a System Context: The Past 80 Years and the Next 20 Years. RR-83-18. International Institute for Applied Systems Analysis, Laxenburg, Austria. July.

Marchetti, C. 1988. Infrastructures for movement: Past and future. In J.H. Ausubel and R. Herman (eds.), Cities and Their Vital Systems: Infrastructure Past, Present, and Future. National Academy Press, Washington, D.C.

Marchetti, C., and N. Nakicenovic. 1979. The Dynamics of Energy Systems and the Logistic Substitution Model. RR-79-13. International Institute for Applied Systems Analysis, Laxenburg, Austria.

Marland, G. 1982. The impact of synthetic fuels on carbon dioxide emissions. In W.C. Clark (ed.), Carbon Dioxide Review. Oxford University Press, New York.

Marland, G., T.A. Boden, R.C. Griffin, S.F. Huang, P. Kanciruk, and T.R. Nelson. 1988. Estimates of CO_2 Emissions from Fossil Fuel Burning and Cement Manufacturing Using the United Nations Energy Statistics and the U.S. Bureau of Mines Cement Manufacturing Data. Oak Ridge National Laboratory, Oak Ridge, Tenn.

Mintzer, I.M. 1987. A Matter of Degrees: The Potential for Controlling the Greenhouse Effect. World Resources Institute, Washington, D.C.

Mintzer, I.M. 1988. Projecting Future Energy Demand in Industrialized Countries: An End-Use Oriented Approach. U.S. Environmental Protection Agency, Washington, D.C.

Myers, N. 1980. Conversion of Tropical Moist Forests. National Academy of Sciences, Washington, D.C.

Myers, N. 1984. The Primary Source: Tropical Forests and Our Future. Norton, New York.

Myers, N. 1986. Tropical Forests: Patterns of Depletion. Tropical Rain Forests and the World Atmosphere. AAAS Select Symposium 101. Westview Press, Boulder, Colo.

Nakicenovic, N. 1988. Dynamics and replacement of U.S. transport infrastructures. In J.H. Ausubel and R. Herman (eds.), Cities and Their Vital Systems: Infrastructure Past, Present, and Future. National Academy Press, Washington, D.C.

National Research Council. 1988. Toward an Understanding of Global Change: Initial Priorities for U.S. Contributions to the International Geosphere-Biosphere Program. National Academy Press, Washington, D.C.

Nordhaus, W.D., and G.W. Yohe. 1983. Paths of Energy and Carbon Dioxide Emissions in Changing Climate. National Academy Press, Washington, D.C.

Nriagu, J.O. 1989. A global assessment of natural sources of atmospheric trace metals. Nature 338:47-49.

Nriagu, J.O., and J.M. Pacyna. 1988. Quantitative assessment of worldwide contamination of air, water and soil by trace metals. Nature 333:134-139.

Palo, M. 1987. Deforestation perspectives for the tropics: A provisional theory with pilot applications. In M. Kallio, D.P. Dykstra, and C.S. Binkley (eds.), The Global Forest Sector. John Wiley and Sons, Chichester, England.

Parikh, K.S. 1981. Exploring National Food Policies in an International Setting. Publication no. WP-81-12. International Institute for Applied Systems Analysis, Laxenburg, Austria.

Parks, P.J. (ed.). 1988. Land Area Modeling and Its Use in Policy: A Workshop on Current Research. Duke University, Durham, N.C.

Parks, P.J., and R.J. Alig. 1988. Land based models for forest resource supply analysis: A critical review. Can. J. For. Res. 18:965-973.

Peterka, V. 1977. Macrodynamics of Technological Change: Market Penetration by New Technologies. RR-77-22. International Institute for Applied Systems Analysis, Laxenburg, Austria. November.

Phipps, T.T., P.R. Crosson, and K.A. Price (eds.). 1986. Agriculture and the Environment. Resources for the Future, Washington, D.C.

Pyne, S.J. 1982. Fire in America, A Cultural History of Wildland and Rural Fire. Princeton University Press, Princeton, N.J.

Radian Corporation. 1988. Emissions and Cost Estimates for Globally Significant Anthropogenic Combustion Sources of NO_x, N_2O, CH_4, CO, and CO_2. U.S. Environmental Protection Agency, Research Triangle Park, N.C.

Richards, J.F. 1986. World environmental history and economic development. In W.C. Clark and R.E. Munn (eds.), Sustainable Development of the Biosphere. Cambridge University Press, New York.

Richards, J.F., J.S. Olson, and R.M. Rotty. 1983. Development of a Data Base for Carbon Dioxide Releases Resulting from Conversion of Land to Agricultural Uses. Institute for Energy Analysis, Oak Ridge, Tenn.

Richards, J.R., and R.P. Tucker. 1988. World Deforestation in the Twentieth Century. Duke University Press, Durham, N.C.

Robinson, J.M. 1987. The Role of Fire on Earth: A Review of the State of Knowledge and a Systems Framework for Satellite and Ground-Based Observations. PhD dissertation. Department of Geography, University of California, Santa Barbara.

Robinson, J.M. 1989. On uncertainty in the computation of global emissions from biomass burning. Climatic Change 14(3):243-261.

Rotty, R.M.. and C.D. Masters. 1985. Carbon dioxide from fossil fuel combustion: Trends, resources and technological implications. In J.R. Trabalka (ed.), Atmospheric Carbon Dioxide and the Global Carbon Cycle. U.S. Department of Energy, Washington, D.C.

Ruddle, K. 1987. The impact of wetland reclamation. In M.G. Wolman and F.G.A. Fournier (eds.), Land Transformation in Agriculture (SCOPE 32). John Wiley and Sons, Chichester, England.

Ruttan, V.W. 1985. Technical and Institutional Change in Agricultural Development: Two Lectures. Economic Development Center, Department of Agricultural and Applied Economics, University of Minnesota, Sanford.

Sathaye, J.A., A.N. Ketoff, L.J. Schipper, and S.M. Lele. 1988. An End-Use Approach to Development of Long-Term Energy Demand Scenarios for Developing Countries. U.S. Environmental Protection Agency, Washington, D.C.

Schurr, S.H. 1984. Energy use, technological change, and productive efficiency: An economic-historical interpretation. In Annual Review of Energy 1984. Annual Reviews Press, Palo Alto, Calif.

Seiler, W., and P.J. Crutzen. 1980. Estimation of gross and net fluxes of carbon between the biosphere and the atmosphere from biomass burning. Climatic Change 2:207-247.

Shah, M.M., G.M. Higgins, A.H. Dassam, and G. Fischer. 1985. Land Resources and Productivity Potential—Agro-ecological Methodology for Agricultural Development Planning. Publication no. CP-85-14. International Institute for Applied Systems Analysis, Laxenburg, Austria.

Shaw, R. 1989. The Study of Chemical Pollution and Its Sources in Dutch Estuaries and Coastal Regions, a Proposal for a Collaborative Agreement. International Institute for Applied Systems Analysis, Laxenburg, Austria.

Spinrad, B.I. 1980. Market Substitution Models and Economic Parameters. RR-80-28. International Institute for Applied Systems Analysis, Laxenburg, Austria. July.

Srinivasan, T.N. 1988. Population growth and food, an assessment of issues, models, and projections. In R. Lee et al. (eds.), Population, Food and Rural Development. Clarendon Press, Oxford, England.

Stavins, R.N. 1989. Alternative Renewable Resource Strategies: A Simulation of Optimal Use. Discussion Paper No. E-89-10. Energy and Environmental Policy Center, Harvard University, Cambridge, Mass.

Stavins, R.N., and A.B. Jaffe. 1988. Forested Wetland Depletion in the United States: An Analysis of Unintended Consequences of Federal Policy and Programs. Discussion Paper No. 1391. Institute of Economic Research, Harvard University, Cambridge, Mass.

Stigliani, W.M., F.M. Brouwer, R.E. Munn, R.W. Shaw, and M. Antonovsky. 1989a. Future Environments for Europe: Some Implications of Alternative Development Paths, Executive Summary. International Institute for Applied Systems Analysis Executive Report 15. IIASA, Laxenburg, Austria.

Stigliani, W.M., F.M. Brouwer, R.E. Munn, R.W. Shaw, and M. Antonovsky. 1989b. Future Environments for Europe: Some Implications of Alternative Development Paths. RR-89-5. International Institute for Applied Systems Analysis, Laxenburg, Austria.

Subak, S. 1989. Accountability for Climate Change. Discussion paper. Kennedy School of Government, Harvard University, Cambridge, Mass.

Svedin, U., and B. Aniansson (eds.) 1987. Surprising Futures, Notes from an International Workshop on Long-term World Development. Swedish Council for Planning and Coordination of Research, Stockholm.

Toth, F.L., E. Hizsnyik, and W.C. Clark (eds.). 1989. Scenarios of Socioeconomic Development for Studies of Global Environmental Change: A Critical Review. RR-89-4. International Institute for Applied Systems Analysis, Laxenburg, Austria.

Tucker, R.P., and J.F. Richards. 1983. Global Deforestation and the Nineteenth Century World Economy. Duke University Press, Durham, N.C.

Turner, B.L., II, W.C. Clark, R.W. Kates, J.T. Mathews, J.R. Richards, and W. Mayer (eds.). 1990. The Earth as Transformed by Human Action. Proceedings of an international symposium held at the Graduate School of Geography, Clark University, Worcester, Mass., October 25-30, 1987. Cambridge University Press, New York.

United Nations. 1985. World population prospects, estimates, and projections as assessed in 1982. Population Studies No. 865. ST/ESA/SER.A/86. United Nations, New York.

U.S. Bureau of the Census. 1987. World Population Profile: 1987. U.S. Department of Commerce, Washington, D.C.

U.S. Department of Energy (DOE). 1988. An Assessment of the Natural Gas Resource Base of the United States. U.S. DOE, Washington, D.C.

U.S. Environmental Protection Agency (EPA). 1987. Assessing the Risks of Trace Gases That Can Modify the Stratosphere. Office of Air and Radiation, U.S. Environmental Protection Agency, Washington, D.C.

U.S. Environmental Protection Agency (EPA). 1988a. Regulatory Impact Analysis: Protection of Stratospheric Ozone. Office of Air and Radiation, U.S. Environmental Protection Agency, Washington, D.C.

U.S. Environmental Protection Agency (EPA). 1988b. Policy Options for Stabilizing Global Climate. Office of Policy, Planning, and Evaluation. DRAFT. February.

Williams, M. 1974. The Making of the South Australian Landscape, a Study in the Historical Geography of Australia. Academic Press, New York.

Williams, R.H., E.D. Larson, and M.H. Ross. 1987. Materials, affluence, and industrial energy use. In Annual Review of Energy 1987. Annual Reviews Press, Palo Alto, Calif.

Wolman, M.G., and F.G.A. Fournier (eds.). 1987. Land Transformation in Agriculture (SCOPE 32). John Wiley and Sons, New York.

Woodwell, G.M., J.E. Hobbie, R.A. Houghton, J.M. Melillo, B. Moore, B.J. Peterson, and G.R. Shaver. 1983. Global deforestation: Contribution to atmospheric carbon dioxide. Science 222:1081-1086.

World Bank. 1984. World Development Report 1984. Oxford University Press, New York.

World Bank. 1987. World Development Report 1987. Oxford University Press, New York.

World Resources Institute (WRI) and International Institute for Environment and Development (IIED). 1988. World Resources 1988-1989. Basic Books, New York.

Zachariah, K.C., and M.T. Vu. 1988. World Population Projections: 1987-88 Edition. Johns Hopkins University Press, Baltimore.

Appendix C
Related Institutional Efforts on Human Interactions with Global Change

OTHER EFFORTS OF THE NATIONAL RESEARCH COUNCIL, THE NATIONAL ACADEMY OF SCIENCES, THE NATIONAL ACADEMY OF ENGINEERING, AND THE INSTITUTE OF MEDICINE

The National Academy of Engineering
(Program Office, NAS 310, National Academy of Engineering, 2101 Constitution Avenue, N.W., Washington, DC 20418)

The National Academy of Engineering (NAE) has embarked on a major program to examine the ways in which innovative technology can be brought to bear on important environmental challenges. This program is organized along three broad themes: (1) *International cooperation and global incentives.* This effort is concerned with policy options that can facilitate progress on transnational environmental issues. (2) *Technological innovation: research and development and applications.* This theme is particularly relevant to the work of the Committee on Global Change. In this area, the NAE intends to further develop frameworks for analyzing the interactions between technology and the environment, and to apply these frameworks to specific industrial sector studies. Emphasis in these studies will be placed on the technology and economics of source reduction (waste minimization) processes. (3) *Economic and institutional implications.* This effort will explore both barriers and possible incentive schemes for explicitly including environmental costs in economic indicators and analysis, and also barriers and incentives for the commercialization of more environmentally compatible technologies. The program has received initial funding of $450,000 through a grant from the Mellon Foundation.

Commission on Behavioral and Social Sciences and Education
(Daniel Druckman or Paul C. Stern, Commission on Behavioral and Social Sciences and Education, National Research Council, 2101 Constitution Avenue, N.W., Washington, DC 20418)

The Commission on Behavioral and Social Sciences and Education at the National Research Council (NRC) will undertake an 18-month study on the human dimensions of global change. The study will be conducted by the 15-member Committee on the Human Dimensions of Global Change, chaired by Oran R. Young of Dartmouth College and directed at the NRC by Daniel Druckman and Paul C. Stern. The overall objective of the study is to develop guidelines for a national social science research program that would contribute to the goals of the International Geosphere-Biosphere Program (IGBP). The committee will also consider issues of collaborative research between the social and natural sciences. Four tasks will be undertaken: (1) an assessment of previous social science research on topics related to global change; (2) an evaluation of extant data resources for social and behavioral research on global change; (3) a consideration of how collaborative research of global change might influence the generation of knowledge in the social sciences as well as attract social and behavioral scientists to apply their knowledge to global issues; and (4) the development of a research agenda that can be implemented over a period of several years. Each of these tasks will address the possibility that social science research can contribute to the international research effort on global change in the natural sciences. The results will also have implications for the future development of the social sciences.

Committee on Science, Engineering and Public Policy
(Robert Coppock, Committee on Science, Engineering and Public Policy, National Academy of Sciences, 2101 Constitution Avenue, N.W., Washington, DC 20418)

The Committee on Science, Engineering and Public Policy will be undertaking a study on the policy implications of greenhouse warming, as described below.[1]

> At the request of Congress, the U.S. Environmental Protection Agency has commissioned a study on policy implications of greenhouse warming by the Committee on Science, Engineering and Public Policy, a unit of the councils of the National Academy of Sciences, the National Academy of Engineering, and the Institute of Medicine. The panel is expected to prepare a report before the end of 1990.
> The study will review research and analysis relevant to greenhouse warming. A careful assessment will be made of existing data and what must be

done to improve current understanding of the underlying phenomena. The report will include, if appropriate, recommendations for actions that would be needed to mitigate and adapt to greenhouse effects if such actions be warranted. It will examine both the underlying phenomena and the expected efficiency and effectiveness of policy interventions in a comprehensive analysis.

A main focus of the study will be on policy interventions and their relative effectiveness. Although the study will be addressed in large part to U.S. policy officials, many of the options to be assessed may require multinational effort. The study is expected to contribute a careful, technically sound review that will be of use to Congress as well as to the Executive Branch and the international community.

The study will be carried out by a panel to synthesize the results and three groups looking at the interrelated issues of direction and rate of change, mitigation policies and their effectiveness, and adaptation strategies.

OTHER DOMESTIC EFFORTS

The Social Science Research Council
(Richard Rockwell, Social Science Research Council,
605 Third Avenue, New York, NY 10158)

The Social Science Research Council (SSRC) has formed the Committee for Research on Global Environmental Change. The program of this committee is given below:[2]

> The committee will foster collaborative interdisciplinary research in six major areas of large-scale, long-term change in the human environment. These areas include land use, global epidemiology, managing global change, security and the environment, sustainable development, and usable knowledge. The committee will advance the state of knowledge in these six areas through the research reviews that it commissions and its synthetic critique of these bodies of research. The committee will sponsor workshops, conferences, intensive working groups, networks, and review articles. Through these six working groups, some 50 researchers will design and execute individual projects that are integrated with those of their colleagues in other disciplines and institutions. The committee and its working groups will seek to construct common definitions, concepts, and methods in an enormous field now lacking such standards and to advance the development of shared data bases.

The SSRC has received initial core support of $127,166 from the National Science Foundation for an 18-month program that began September 1989.

The working group on land use has a research agenda that overlaps substantially with the agenda outlined by the CGC in chapter 4. It is important to recognize that while the SSRC will play a critical role in furthering this agenda through the program described above, it will not be funding investigator-initiated research. The tentative plans of the working group on land use are outlined below.

Objective: (1) to develop an understanding of the associations among the driving and mitigating forces of land-use change, proximate sources of this change, and the land use change/land cover connection; (2) to "test" for the variable dimensions of change by spatial scale and situation; (3) to create likely scenarios of futures of selected land-use changes and environmental impacts.

Specific objectives are the following:

1. Refine and elaborate global land-use classification system.

2. Identify four to five major classes in this system that are particularly significant to global environmental change and will undergo major change in the near future, some of which can be matched to forest and wetlands.

3. Identify the proximate sources of change for each land use, initiating case studies of the driving and mitigating forces that lead to the states of and rates of change in each proximate source. Presumably four of these to match those identified by HIB.

4. Integrate items 1 to 3 to create global and regional tests of the roles of each force in the aggregate and by circumstance.

5. Create scenarios of short- and long-term changes in the land uses from items 1 to 4 and associated with impacts on land cover.

The National Science Foundation
(Roberta Balstad Miller, Division of Social and Economic Science, National Science Foundation, 1800 G Street, N.W., Washington, DC 20550)

The National Science Foundation (NSF) has a research program on the Human Dimensions of Global Environmental Change. The call for proposals describes this program as follows:

> Recent interest in processes of global environmental change has led biological and geoscientists to undertake major new research efforts in the United States and elsewhere. A number of workshops and conferences have recently stressed that these inquiries into natural processes of change must be complemented by social science investigations to understand how human activity affects and is affected by global environmental change. To encourage research in this broad area, the Division of Social and Economic Science welcomes proposals for research on the human dimensions of global environmental change. These dimensions include but are not limited to such broad topics as the social, economic, demographic, governmental, and institutional components of global change. Studies of human influences on the environment and institutional responses to global changes are both appropriate for this initiative, but proposals must emphasize fundamental research into processes of change over time or space.

Awards for the fiscal year 1989 totaled $750,000. Funding was awarded to four investigator-initiated research projects. The other six awards were for conferences, workshops, or support for institutionally based committees that were developing research agendas or research programs on global environmental change.

OTHER INTERNATIONAL EFFORTS

The Human Dimensions of Global Change, An International Programme on Human Interactions with the Earth
(Peter Timmerman, Human Dimensions of Global Change, Interim Secretariat, c/o IFIAS, 39 Spadina Road, Toronto, Canada M5R 2S9)

The current steering committee for this program consists of the International Federation of Institutes for Advanced Study (IFIAS), The International Social Science Council (ISSC), the United Nations University (UNU), and the United Nations Educational, Scientific, and Cultural Organization (UNESCO). The objectives of the research program are as follows:[3]

- to improve scientific understanding and increase awareness of the complex dynamics governing human interaction with the total Earth system;
- to strengthen efforts to study, explore, and anticipate social change affecting the global environment;
- to identify broad social strategies to prevent or mitigate undesirable impacts of global change, or to adapt to changes that are already unavoidable;
- to analyze policy options for dealing with global environmental change and promoting the goal of sustainable development.

International Social Science Council
(Harold Jacobson, Center for Political Studies/ISR, University of Michigan, P.O. Box 1248, Ann Arbor, MI 48106-1248)

The International Social Science Council's (ISSC) program is described in their recent document "Plan of Action for Research on the Human Dimensions of Global Environmental Change."[4] This program recognizes three broadly defined research areas: "the social dimensions of resource use; the perception and assessment of global environmental conditions and change; and the impacts of local, national, and international social, economic, and political structures and institutions on the global environment." Within these broad areas, seven research topics are outlined in the plan:

1. social dimensions of resource use; 2. perception and assessment of global environmental conditions and change; 3. impacts of local, national and

international social, economical and political structures and institutions; 4. land use; 5. energy production and consumption; 6. industrial growth; and 7. environmental security and sustainable development.

European Science Foundation
(Timothy O'Riordan, School of Environmental Sciences, University of East Anglia, Norwich, England NR4 7TJ)

At its 1988 General Assembly, the European Science Foundation established a planning committee on environment and development. This committee is preparing a detailed research agenda in two areas: (1) environmental economics, with an emphasis on resource auditing, and (2) the theories and issues behind institutional adaptation to environmental change.

International Institute for Applied Systems Analysis
(Robert H. Pry, Director, IIASA, A-2361 Laxenburg, Austria; Peter de Janosi, U.S. IIASA Council Member, Russell Sage Foundation, 112 East 64th Street, New York, NY 10021; Alan McDonald, Staff Director, U.S. Committee for IIASA, American Academy of Arts and Sciences, 136 Irving Street, Cambridge, MA 02138)

The International Institute for Applied Systems Analysis (IIASA) conducts research in five program areas, four of which undertake applied research that is relevant to human interactions with global environmental change. These programs are Environment; Population; Technology, Economy and Society (TES); and Climate and Ecology Related Energy. IIASA is also the hub of an international network using the Basic Linked System (BLS) of agricultural models developed at the institute between 1976 and 1985.

The Environment, Population, and TES programs and the agricultural project maintain a number of data sets and perform research that looks both at the human sources of global change and the impacts of global change. The newest of IIASA's four applied programs, the Climate and Ecology Related Energy Program, was established in 1990 to integrate a number of the ongoing lines of research at the institute, and to sharpen their focus on the connection between patterns of energy technology, use, and emissions on the one hand, and international environmental changes on the other.

Research and data sets related to human sources of global change include:

- International population growth, migration, and demographics;
- International agricultural production and fertilizer use;
- Consumption, efficiency, and emissions associated with different energy technologies;

- International patterns, and projections, of energy production, consumption, and emissions;
- Regional strategies for increasing energy efficiency and reducing emissions;
- Long-term regularities in technological and industrial evolution, diffusion, and replacement, with particular expertise in the energy and transportation sectors;
- European emission, transport, and deposition of sulfur oxides, nitrogen oxides, photochemical oxidants, and heavy metals; and
- European land use and forest resources.

Data sets and research activities related to the impacts of global change include:

- Climate change impacts on regional agriculture and on international agricultural trade;
- Climate change impacts on global vegetation patterns;
- Climate change impacts on the boreal forests;
- European soil and lake acidification;
- International soil loss;
- European toxic waste accumulations; and
- Regional impacts of pollution on water resources, air quality, and public health.

NOTES

1. National Academy of Sciences, National Academy of Engineering, Institute of Medicine. 1989. "NAS/NAE/IOM Study on Policy Implications of Greenhouse Warming." Proposal to EPA.

2. Social Science Research Council Annual Report (draft), 1988-1989.

3. "The Human Dimensions of Global Change, An International Programme on Human Interactions with the Earth, Proposed Programme." Draft document prepared for the informal consultation of donor organizations, Ottawa, March 22-23, 1989.

4. International Social Science Council. "Plan of Action for Research on the Human Dimensions of Global Environmental Change." Draft 4.2. December 9, 1989.